Statistics for Innovation

Pasquale Erto
Editor

Statistics for Innovation

Statistical Design
of "Continuous" Product Innovation

 Springer

Pasquale Erto
Department of Aerospace Engineering
University of Naples "Federico II", Italy

ISBN 978-88-470-0814-4   ISBN 978-88-470-0815-1 (eBook)

DOI 10.1007/978-88-470-0815-1

Library of Congress Control Number: 2008939992

Cover concept: Simona Colombo, Milano
Cover picture: Riccardo Dalisi
Typesetting: le-tex publishing services oHG, Leipzig, Germany

Printed on acid-free paper

Springer-Verlag Italia – Via Decembrio 28 – 20137 Milano

springer.com

Foreword and Acknowledgements

This book collects the main contributions to the research program "Statistical design of the *continuous* product innovation" carried out by a group of five units from the Italian Universities of Naples Federico II, Bologna, Salerno, Palermo and the Polytechnic of Turin, respectively. Some contributions are mainly applicative and concern the innovation of specific products or processes; other contributions are mainly theoretical and propose new methods to support the innovation process. All ones are aimed at showing that the technological innovation can be planned and designed likely any other product feature.

Until the past century, firms gained competitive advantage from reducing costs, minimizing process variability and continuously improving their products. In the current global marketplace, these initiatives are only prerequisites to survive, while innovation is the actual source of competitive advantage.

Product innovation is realized when the product is provided with a new feature aimed at fulfilling a new customer need. Since customer needs change rapidly over time, product innovation must be a continuous process and, consequently, it cannot rely on a brilliant and contingent intuition as for an invention. Often some needs are latent and cannot be elicited from a traditional customer survey, so they must be identified differently, by means of more advanced statistical tools. Sometimes, the innovation is implemented by exploiting a new technology, which is not necessarily the most advanced one, since its aim is fulfilling a new customer need not showing an intrinsic novelty.

The above essential considerations are only examples of those developed by the contributions contained in this book. From the whole set of contributions a new approach to the innovation as a never-ending series of manufacturing cycles arises. Each cycle requires a series of steps, each performed using a specific statistical and/or engineering tool.

The contributions result from different Statistics and Engineering schools, hence their whole set is not biased by any specific point of view. Different opinions that were not able to converge were left unchanged and, when possible, all were tested facing practical applications. Nevertheless, many unsolved questions arise, but these certainly will boost further research to advance the statistical and engineering knowledge addressed to innovation.

A glance to the table of contents is the most effective way to master the whole range of practical and theoretical topics covered by this book. It starts from product innovation attained by engineering design in virtual reality; then shows examples of process innovation obtained integrating physical and simulation experiments; proposes a Bayesian approach to the innovation of the reliability and maintenance service; ends with an advanced approach to management of the research and innovation activities themselves.

This book and the whole research program "Statistical design of the *continuous* product innovation" (PRIN 2005–2007) have been financially supported by the Italian Ministry of University and Scientific and Technological Research (MIUR).

Naples, 24 August 2008   *Pasquale Erto*

Contents

Part II Technological Process Innovation

**5 Design for Computer Experiments:
    Comparing and Generating Designs in Kriging Models**   91

Giovanni Pistone and Grazia Vicario

**6 New Sampling Procedures in Coordinate Metrology
    Based on Kriging-Based Adaptive Designs** 103

Paola Pedone, Daniele Romano, and Grazia Vicario

**7 Product and Process Innovation by Integrating Physical
    and Simulation Experiments** 123

Daniele Romano

Part IV Research and Innovation Management

List of Contributors

Baldi Antognini, Alessandro
Department of Statistical Sciences, University of Bologna
e-mail: a.baldi@unibo.it

Barone, Stefano
Department of Technology, Production and Managerial Engineering,
University of Palermo, Viale delle Scienze, 90128 Palermo
e-mail: stefano.barone@unipa.it

Carfagna, Elisabetta
Department of Statistical Sciences, University of Bologna
e-mail: elisabetta.carfagna@unibo.it

De Chiara, Gaetano
AVIO S.p.A., Manufacturing Technologies Department, Pomigliano, Naples, Italy
e-mail: gaetano.dechiara@aviogroup.com

Erto, Pasquale
Department of Aerospace Engineering, University of Naples Federico II,
P. le Tecchio 80, 80125, Naples, Italy
e-mail: pasquale.erto@unina.it

Franceschini, Fiorenzo
Department of Production Systems and Business Economics (DISPEA),
Politecnico di Torino, Corso Duca degli Abruzzi 24, 10129, Torino, Italy
e-mail: fiorenzo.franceschini@polito.it

Galetto, Maurizio
Department of Production Systems and Business Economics (DISPEA),
Politecnico di Torino, Corso Duca degli Abruzzi 24, 10129, Torino, Italy
e-mail: maurizio.galetto@polito.it

Giorgio, Massimiliano
Department of Aerospatial and Mechanical Engineering,
Second University of Naples, 81031 Aversa (NA), Italy
e-mail: massimiliano.giorgio@unina2.it

Giovagnoli, Alessandra
Department of Statistical Sciences, University of Bologna
e-mail: alessandra.giovagnoli@unibo.it

Guida, Maurizio
Department of Information Engineering and Electrical Engineering,
University of Salerno, 84084 Fisciano (SA), Italy
e-mail: mguida@unisa.it

Lanzotti, Antonio
Department of Aerospace Engineering, University of Naples Federico II,
P. le Tecchio 80, 80125 Naples, Italy
e-mail: antonio.lanzotti@unina.it

Lombardo, Alberto
Department of Technology, Production and Managerial Engineering,
University of Palermo, Viale delle Scienze, 90128 Palermo
e-mail: alberto.lombardo@unipa.it

Maisano, Domenico A.
Department of Production Systems and Business Economics (DISPEA),
Politecnico di Torino, Corso Duca degli Abruzzi 24, 10129, Torino, Italy
e-mail: domenico.maisano@polito.it

Marrone, Roberto
AVIO S.p.A., Manufacturing Technologies Department, Pomigliano, Naples, Italy
e-mail: roberto.marrone@aviogroup.com

Marzialetti, Johnny
Department of Statistical Sciences, University of Bologna
e-mail: johnny.marzialetti@unibo.it

Mastrogiacomo, Luca
Department of Production Systems and Business Economics (DISPEA),
Politecnico di Torino, Corso Duca degli Abruzzi 24, 10129, Torino, Italy
e-mail: luca.mastrogiacomo@polito.it

Matrone, Giovanna
Department of Aerospace Engineering, University of Naples Federico II,
P. le Tecchio 80, 80125 Naples, Italy
e-mail: giovanna.matrone@unina.it

Pallotta, Giuliana
Department of Aerospace Engineering, University of Naples Federico II,
P. le Tecchio 80, 80125, Naples, Italy
e-mail: g.pallotta@unina.it

Palumbo, Biagio
Department of Aerospace Engineering, University of Naples Federico II,
P. le Tecchio 80, 80125, Naples, Italy
e-mail: biagio.palumbo@unina.it

Pedone, Paola
INRIM (Italian National Research Institute of Metrology),
Strada delle Cacce, 91, 10135 Torino, Italy
e-mail: p.pedone@inrim.it

Pistone, Giovanni
Politecnico di Torino, Department of Mathematics,
Corso Duca degli Abruzzi, 24, 10129 Torino, Italy
e-mail: giovanni.pistone@polito.it

Pulcini, Gianpaolo
Istituto Motori, National Research Council, 80125 Napoli, Italy
e-mail: g.pulcini@im.cnr.it

Romano, Daniele
Department of Mechanical Engineering, University of Cagliari,
Piazza d'Armi, 1, 09123 Cagliari, Italy
e-mail: romano@dimeca.unica.it

Tarantino, Pietro
Department of Aerospace Engineering, University of Naples Federico II,
P. le Tecchio 80, 80125 Naples, Italy
e-mail: pietro.tarantino@unina.it

Vanacore, Amalia
Department of Aerospace Engineering, University of Naples Federico II,
P. le Tecchio 80, 80125 Naples, Italy
e-mail: amalia.vanacore@unina.it

Vicario, Grazia
Politecnico di Torino, Department of Mathematics,
Corso Duca degli Abruzzi, 24, 10129 Torino, Italy
e-mail: grazia.vicario@polito.it

Zagoraiou, Maroussa
Department of Statistical Sciences, University of Bologna
e-mail: maroussa.zagoraiou@unibo.it

Part I
Design for Innovation

Chapter 1
Analysis of User Needs for the Redesign of a Postural Seat System

Stefano Barone, Alberto Lombardo, and Pietro Tarantino

Abstract The identification and translation of customer needs early in the design process is a major challenge for product design researchers. Some needs are explicit and customers can state them very clearly. Other needs are implicit, so customers cannot express them, e.g., those pertaining to the affective and emotional sphere. In this work, we describe the methods most commonly used to capture explicit and emotional customer needs, and the traditional ways in which they are used. Moreover, an integration of QFD and Kansei engineering, a simplification of Kano methodology, and a new attribute weighing methodology based on the "choice time" are discussed for the design of an innovative postural seat system for patients affected by mental retardation.

1.1 Introduction

In recent years, customer-oriented product development has become vital for companies facing global competition. The identification and, above all, the translation of customers' needs early in the design process is a major challenge for product design researchers. These needs have three main characteristics that create difficulties in product development tasks. Firstly, not all of the customers' needs are explicitly

Stefano Barone
Department of Technology, Production and Managerial Engineering
University of Palermo, Viale delle Scienze, 90128 Palermo
e-mail: stefano.barone@unipa.it

Alberto Lombardo
Department of Technology, Production and Managerial Engineering
University of Palermo, Viale delle Scienze, 90128 Palermo
e-mail: alberto.lombardo@unipa.it

Pietro Tarantino
Department of Aerospace Engineering, University of Naples Federico II
P. le Tecchio 80, 80125 Naples, Italy
e-mail: pietro.tarantino@unina.it

or clearly stated by them. Secondly, not all of the customers' needs are easily transformed into engineering characteristics. Thirdly, these needs quickly change due to environmental factors such as advertising. Moreover, while customers expected functionality, reliability and safety from products in the past, these aspects are now increasingly taken for granted. Indeed, the product's affective and emotional properties (or "Kansei" in Japanese) have recently emerged as important factors in the successful marketing of products.

Therefore, methods for eliciting and analyzing customer needs can be successful only if they make use of a multidisciplinary approach in which engineering competences are merged with statistical models, quality tools and psychology concepts. Moreover, the use of a multidisciplinary approach is the solution to the so-called "crisis of the engineering algorithm" (Keniston 1996).

This work aims to demonstrate the advantages of such a multidisciplinary approach using a case study in which an innovative postural seat system for patients affected by mental retardation is designed. The difficulties inherent in this study, as well as the high number of "potential customers," show the validity and usefulness of the proposed product design approach.

The paper is organized as follows. Section 1.2 describes the evolution of the customers' concept of quality and the corresponding evolution in product development strategy. Section 1.3 briefly describe the methodologies used to capture customers' needs and translate them into engineering characteristics. Section 1.4 formalizes the necessary modifications of some of these methods in order to take into account emotional or implicit customer needs. Section 1.5 presents the results from the first part of the case study on the postural seat system design. The last section is reserved for the conclusion of this study and some reflections.

1.2 Evolution in the Customers' Concept of Quality

Over the last few decades, quality has become the leading issue in many companies and other organizations in order to improve competitiveness and increase customer satisfaction (Dahlgaard et al. 2002). Nevertheless, the concept of quality has evolved a great deal, from the product's *conformance to specifications and requirements* (Crosby 1979), to *the product's ability to satisfy the needs and expectations of the customer* (Bergam and Klefsjö 1994). This evolution in the customers' concept of quality has profoundly affected the product design strategy used by designers and engineers (Fig. 1.1). In the age of mass production up to the 1960s, manufacturers designed products according to their own ideas, and then tried to sell them. This product design strategy can be termed the "product-out strategy," and it implied a lack of communication with the customer. At the end of 1959, Deming (1986), during his lecture to top Japanese managers, introduced his approach to designing and producing a product. It was an iterative approach in which customer research was included in order to establish the continuous integration of customers into the product design or redesign process.

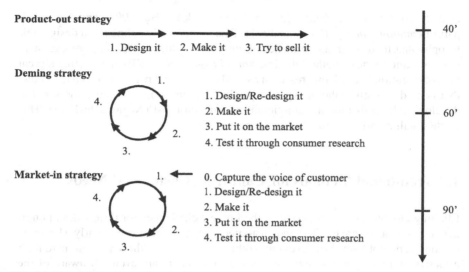

Product-out strategy

1. Design it   2. Make it   3. Try to sell it

40'

Deming strategy

1.
4.
2.
3.

1. Design/Re-design it
2. Make it
3. Put it on the market
4. Test it through consumer research

60'

Market-in strategy

1.
4.
2.
3.

0. Capture the voice of customer
1. Design/Re-design it
2. Make it
3. Put it on the market
4. Test it through consumer research

90'

Fig. 1.1 The evolution in product design strategy

Even though the new approach included the customers' views and enforced continuous improvement, it was not entirely feasible in a new product development context. Companies can no longer afford to have customers evaluate products after they have been launched onto the market. Instead, it is important to build customer satisfaction into the products before their introduction into the market. This product design strategy is termed the "market-in strategy," and it presupposes a great deal of communication with customers.

This communication creates a need to establish efficient means for understanding and integrating customer needs as early as possible in the product development process, and for translating those needs into product characteristics. Many methodologies were originated for these aims. *voice of the customer*, *quality function deployment* (QFD), and the *Kano model* are the tools most commonly used in this context, and they will be briefly described in the next section.

The problem with these methodologies is that they are able to capture and integrate only the conscious and explicit needs of customers. Nowadays, inexplicit and emotional properties are known to be just as important, and so are included as evaluation criteria in the design process, preferably in the early stages. To support this idea, it is necessary to observe that new products introduced by organizations operating in many market sectors are often not as successful as expected, even though they are functionally reliable and produced to high quality standards. This occurs because designers and engineers do not seem to understand the feelings of customers toward the product concept. Different methodologies have been developed and integrated into product design processes in order to measure the affective impacts of different products on customers. Some of these methodologies have been termed *affective design* (Khalid and Helander 2004), *human-centered design* (Toft et al. 2003) and *affective human factors design* (Park and Han 2004), and they are all part of *emotional design* (Norman 2004). This is succinctly defined as a design philosophy that focuses on the influence of emotions on the way humans interact with objects. Among these methodologies, *Kansei engineering* (KE) is attracting a great deal of attention in academic research as well as industrial research. It is a technique that is used to analyze the unexpressed and unconscious needs of customers and to develop such needs into an "emotional" specification list (Nagamachi 1995). This method will be briefly described in the next section.

1.3 Traditional Methods for Capturing Customers' Needs

Eliciting customers' needs is one of the biggest challenges for designer and engineers. Some needs are explicit and customers can state them very clearly. However, customers do not know how to express other needs, such as those pertaining to their affective and emotional sphere. Sometimes, customers are even not aware of the existence of these needs. In this section, we describe the methods most commonly used to capture explicit and emotional customers' needs, and the traditional ways in which they are used.

1.3.1 Voice of the Customer

The voice of the customer (VoC) is a general term for a structured list of customer needs for the product or service being designed (Griffin and Hauser 1993). This list is gathered by asking individual customers or focus groups to talk freely about their needs for the product or service in a survey. The result of the interview is a set of words and phrases representing the customers' wants and needs. These phrases are usually sorted by the *voice of the customer table* (VOCT) (Cohen 1995). The VOCT traditionally has two parts. Part 1 contains information on the source of the customer phrases and on the ways that customers can come into contact with the product/service being designed. In part 2, the data are sorted in different ways according to different categories. The most commonly used categories are customer needs (statement in the customer's words), substitute quality characteristics (SQCs) (statement in the company's technical language) and functions (descriptions of the ways in which the product or service operates). Another tool for sorting and organizing the collected data during the interview is an affinity diagram (Tague 2004). This is a method that is useful for gathering large amounts of data (opinions, ideas, etc.) and for organizing them into groupings based on their relationship. The voice of the customer can be also collected through customer complaints. In particular, the critical incident technique provides a tool for identifying significant factors that contribute to the success or failure of an action (Flanagan 1954). Critical incidents are usually gathered by allowing customers to freely converse about their experiences.

1.3.2 Quality Function Deployment

Quality function deployment (QFD) is a customer-oriented approach to product innovation. It provides a systematic process for translating customer requirements into technical requirements at each stage of product development and production (Sullivan 1986b). Quality function deployment was first successfully used in the 1960s by Japanese manufacturers in the areas of tire production and electronics (Akao and Mazur 2003). The first publication was due to Akao, who was the first to formalize the term "hinshitsu tenkai" (quality deployment) as a method of deploying the main engineering characteristics in order to ensure that quality is incorporated into the design process (Akao 1972). More than 20 years later, Clausing first used the QFD approach in the United States, in the Ford Motor Corporation (Hauser and Clausing 1988).

QFD is a *process* that can help companies to make key trade-offs between what the customer wants and what the company can afford to build (Govers 1996). QFD decomposes the product development process into four phases: strategy and concept definition, product design, process design and manufacturing operations. In each phase, the customer requirements (the "whats") serve as input to establish the engineering characteristics (the "hows") of the product design. The relationships between the inputs and outputs are mapped into matrices (Cohen 1995). The ini-

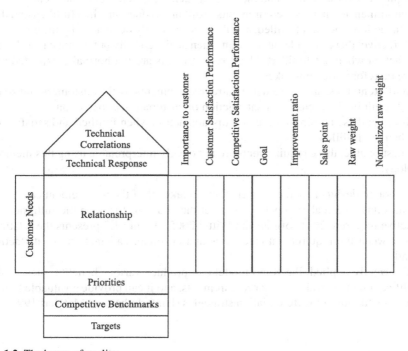

Fig. 1.2 The house of quality

tial and most important matrix, linking the voice of the customer to the engineering characteristics, is the "house of quality" (HOQ). The house of quality procedure can be divided into several steps (Chan and Wu 2005), all of them constituting a section in the house of quality diagram (Fig. 1.2).

If correctly applied, QFD can produce benefits such as a deeper understanding of customer requirements, fewer start-up problems, and fewer design changes which are also made earlier (Lockamy and Khurana 1995).

1.3.3 Kano Model

Developed in the 1980s by Prof. Noriaki Kano, this model aims to understand the relationship between the fulfilment (or not) of a requirement (product feature) and the satisfaction or dissatisfaction experienced by the customer (Kano et al. 1984). In his model, Kano classifies the customer requirements into six categories (CQM 1993).

- Must-be: these are considered prerequisites by the customers. If these requirements are not fulfilled, the customer will be extremely dissatisfied. On the other hand, their fulfilment will not increase his/her satisfaction. Customers take these requirements for granted and therefore does not explicitly demand them.
- One-dimensional: these requirements result in satisfaction when fulfilled and dissatisfaction when not fulfilled, and they are explicitly demanded by the customer.
- Attractive: these provide satisfaction when achieved fully but do not cause dissatisfaction when not fulfilled. These requirements are not normally expected and are therefore often unspoken.
- Indifferent: these are viewed as neutral requirements by the customers and so do not result in either customer satisfaction or customer dissatisfaction.
- Reverse: these requirements cause dissatisfaction when fulfilled and satisfaction when not fulfilled.
- Questionable: these requirements are not clearly interpretable using this methodology.

Even though designers and engineers have to take all of the requirement categories into account, in a real, competitive market they have to focus on the fulfilment of attractive requirements (Lofgren and Wittel 2005). Figure 1.3 presents the relationship between the requirement categories and customer satisfaction/dissatisfaction visually.

If correctly applied, the Kano model can produce various benefits, such as the identification of critical customer requirements, and it can provide a valuable tool in trade-off situations or differentiation strategies (Hinterhuber and Matzlerl 1998).

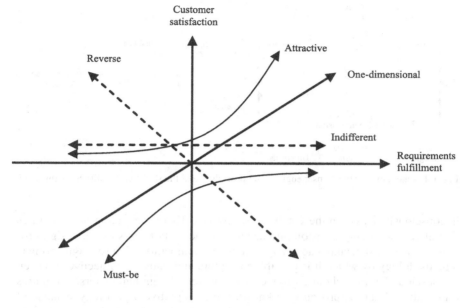

Fig. 1.3 Diagram of the Kano model

1.3.4 Kansei Engineering

The roots of Kansei engineering (KE) can be traced back to the Faculty of Engineering at Hiroshima University, where Professor Nagamachi was appointed to the Engineering Management group in the early 1970s, with the aim of developing emotional ergonomics for product design (Schütte 2005). After several studies of different products such as houses, automobiles, and electrical appliances, he formalized the concept of Kansei engineering as a consumer-oriented technology for new product development which made it possible to translate consumers' feelings about and images of a product into design elements and product features (Nagamachi 1995). The Japanese word "Kansei" is an expression that is not readily translated into other languages because it is very closely connected to the Japanese culture. It typically consists of two different Kanji-signs, "Kan" and "Sei," which in combination mean sensitivity or sensibility (Lee et al. 2002). According to Nagamachi (Nagamachi and Matsubara 1997), *Kansei is the impression somebody gets from a certain artefact, environment or situation using all her/his senses of sight, hearing, feeling, smell, taste as well as their recognition.* For example, we can imagine the situation in which a potential customer wants to buy a car and he/she initially takes it for a test drive. During this test he/she can smell the odor inside the new car, examine the surface in great detail, feel the sound of the motor, and watch the speedometer needle climb. Especially in the new global market, where so many products with the same functionality and quality are available, many customers make their final decisions

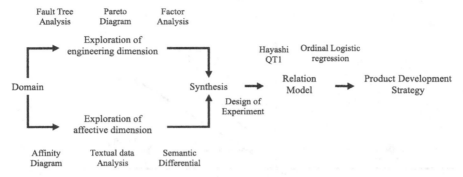

Fig. 1.4 Schematic for the Kansei engineering procedure and the statistical/quality tools involved

unconsciously, based on these subjective feelings. Therefore, taking these feelings that affect the buying decision into account in the early phases of the design process can yield a substantial advantage over one's competitors. Kansei engineering is a methodology by which it is possible to capture and translate subjective and even unconscious feelings about a product into concrete design parameters. It requires a multidisciplinary approach with knowledge of cognitive psychology, behavioral science, psychometrics, consumer research and marketing science (Lanzotti and Tarantino 2007).

To obtain relationships between the customers' Kansei and design parameters, a systematic procedure can be followed. This is shown schematically in Fig. 1.4, where some of the statistical and quality tools that can support the flow of analysis in the procedure are also illustrated (Fonti and Tarantino 2006).

The main idea behind the methodology is to describe the product from two different perspectives, the emotional perspective and the technical perspective. Words and phrases that describe the emotional sphere of customers are linked to the engineering sphere in the synthesis phase, where product concepts are evaluated in a interview session. Data extracted from the synthesis phase constitute the input for the relation model that indicates how the emotional sphere and the engineering sphere are related and the strength of this relationship (Schütte and Eklund 2005). The results of a Kansei engineering procedure allow companies to implement the right product development strategy based on the specific needs and feelings of customers.

1.4 Advanced Methods for Capturing Customers' Needs

Traditionally, methods for capturing customers' needs were often applied individually, in order to capture either declared and explicit needs or emotional and implicit needs. The integration of these methods was only performed for the first task (integration of VOC with QFD and QFD with Kano model). Given the evidence that the emotional properties of a product or service must be considered early on in the design phase, an integration of QFD and Kansei engineering methodology is proposed here. Moreover, there is a lack of methods which can capture customer needs that cannot be expressed in words. In the second part of this section, we propose a method for capturing customer preferences for product attributes that uses an indirect value, such as the time he/she takes to choose a ranking in a controlled interview. The last part of the section describes a modification of the Kano methodology that allows similar information to be obtained with a simplified version of Kano's questionnaire.

1.4.1 Integrating Kansei Engineering

Even though, according to Mazur (1997), QFD makes it possible to translate both spoken and unspoken needs into engineering characteristics, this methodology has been always used to translate declared and explicit customer needs. On the other hand, Kansei engineering aims to identify the emotional needs of the customers and the relations these have to the technical aspects of design. Therefore, QFD and Kansei engineering have the same goal but they use different data. Since both methodologies employ a systematic step-by-step approach, a strategy in which the results of the two methodologies are merged somehow is feasible. In particular, the more general structure of QFD can be integrated with the results of a simplified Kansei engineering approach. By a "simplified Kansei engineering approach," we mean a process in which the links between Kansei words and engineering characteristics are primarily explored using qualitative tools (an example of a simplified Kansei engineering approach can be found in Lanzotti and Tarantino 2007).

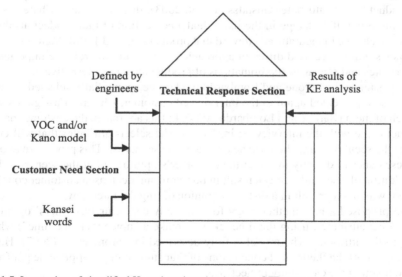

Fig. 1.5 Integration of simplified Kansei engineering into the QFD process

The integration of QFD and Kansei engineering can take place in the *customer needs* and *technical response* sections of the HOQ. The *customer needs* section of the HOQ, the core of a QFD approach, uses the data resulting from the application of the VOC method or/and the Kano model. This section can be divided into two subsections. The first takes into account explicit and declared customers needs, and the second considers the emotional needs expressed by the *Kansei words*. The *technical response* section defines one or a few technical performance measurements for each customer need. Again, this section can be divided into two subsections. The first takes into account the engineering characteristics corresponding to declared needs (often defined by engineers), and the second considers the technical properties linked with Kansei words, which arise from the simplified Kansei engineering approach.

A first tentative theory for integrating QFD and Kansei engineering can be found in Arnold (2001), and is presented visually in Fig. 1.5. The study presented in Sect. 1.5 uses the practical integration of these methodologies as central methodology.

1.4.2 A New Practical Way to Measure Customer Preferences for Product Attributes

Huge amounts of time and effort have been expended by consumer researchers in order to develop methods for identifying product attributes that are important in influencing product preferences and choice. Among these methods, conjoint analysis has been broadly used to estimate the value that customers associate with a particular product feature/attribute (Gustafsson et al. 2003). In general, an attribute is said to be important if a change in the individual's perception of that product attribute leads to a change in their attitude toward that product (Jaccard 1996). Many conjoint analysis studies have used different approaches to measure the relative importance of attributes and scenarios (combinations of product attribute alternatives).

For instance, in Barone and Lombardo model, the sequentially selected scenarios are then presented again to the interviewed customer, who must assign a score to each of them (Barone and Lombardo 2004). In this approach, the customer has to interact twice with the interviewer; initially he/she selects attributes that will constitute the scenarios, and then he/she evaluates the scenarios. This procedure allows the researcher to directly interpret the customer's opinion about the scenarios, but the addition of a second step can result in boredom and decreased customer concentration, which may result in a distorted opinion during the interview.

We proposed a new methodology for indirectly capturing customers' opinions on product attributes, using the time taken to make a choice ("choice time") when ranking the attributes. The model is fully described in Barone et al. (2007). Here, we only report the theoretical conclusions of that study and the applicable platform for conducting the case study in Sect. 1.5.

The weights for each attribute $(0 \leq w_i \leq 1)$ are calculated by solving the system of $n+1$ equations:

$$\begin{cases} \dfrac{w_i}{w_{i+1}} = 1 + \dfrac{t^*}{t_{\mathrm{c}}^{(i)}} & i = 1, 2, \ldots, n-1 \\ \sum\limits_{i=1}^{n} w_i = 1 \end{cases} \tag{1.1}$$

where $t_{\mathrm{c}}^{(i)}$ is the time a respondent takes to choose the position i-th in the ranking process (choice time) and t^* is a reference time testing the degree to which the respondent reacts to a predefined stimulus (reaction time). System solutions can be seen as applications of a recursive calculus, by posing $1 + \dfrac{t^*}{t_{\mathrm{c}}^{(i)}} = a_i$. Then the weights of importance are:

$$w_i = \frac{a_i \times a_{i+1} \times \ldots \times a_{n-1}}{(a_1 \times a_2 \times \ldots \times a_{n-1}) + (a_2 \times a_3 \times \ldots \times a_{n-1}) + \ldots + (a_{n-3} \times a_{n-2} \times a_{n-1}) + a_{n-1} + 1} \tag{1.2}$$

for $i = 1, 2, \ldots, n-1$ and

$$w_n = \frac{1}{(a_1 \times a_2 \times \ldots \times a_{n-1}) + (a_2 \times a_3 \times \ldots \times a_{n-1}) + \ldots + (a_{n-3} \times a_{n-2} \times a_{n-1}) + a_{n-1} + 1} \tag{1.3}$$

These recursive formulae were implemented with a JAVA code, and a software interface (EAW[1] – Easy Attribute Weighing) was defined, which was very useful for the interview task. This interface allows the experimenter to automate the customers' attribute ranking process and the weight calculation process. Moreover, it provides a way to directly rate the attributes such that the results of the two procedures can be prepared. A visual report, together with an EXCEL file, will be generated that contains the ranks of the attributes, the choice times, the weights, the rating scores and the t^* values for each respondent. EAW is a flexible tool for applying this methodology in different experimental contexts and with several customers.

By measuring the respondent choice time and using this methodology, it is possible to extrapolate from each respondent not only the preference order of attributes but also the importance (the weight) of each attribute.

1.4.3 A Simplified Version of the Kano Questionnaire

The traditional methodology for mapping customer needs into the Kano model makes use of a questionnaire. In its standard form, the questionnaire consists of two questions for each customer requirement: a functional question captures the customer's feelings when the requirement is fulfilled, and a dysfunctional question

---

[1] License released under the authors' authorization

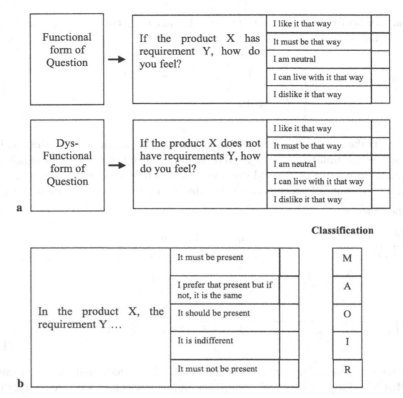

a

Classification

In the product X, the requirement Y ...	It must be present		M
	I prefer that present but if not, it is the same		A
	It should be present		O
	It is indifferent		I
	It must not be present		R

b

Fig. 1.6a,b Kano questionnaire: **a** traditional version; **b** simplified version

captures the customer's feelings when the requirement is not fulfilled (see Fig. 1.6a). By combining the two answers in the Kano evaluation table, the customer requirements can be classified according to the categories defined above. Therefore, the traditional methodology is divided into two steps: (1) collecting data in a questionnaire, often quite a time-consuming task, and (2) combining the data in a predefined table. Due to this elaborate process, the risk of bias during data analysis is high. Moreover, based on our past experiences, many respondents have found the "double-question" format to be rather contradictory.

To simplify the task of the respondent and the analysis, we propose a questionnaire with a single question for each customer requirement. The chosen form is that of Fig. 1.6b, and this allows a clear interpretation of the examined requirement. The interview process is not influenced by biased answers, and the methodology is reduced to one step.

1.5 Needs Analysis for the Design of a Postural Seat System

This study aims to develop an innovative postural seat system for children affected by mental retardation. The results that will be described here and in the next chapter of this book were obtained in close collaboration with IRCSS Oasi Maria SS., a research institution of internationally recognized excellence in the area of mental retardation and brain aging, located in the center of Sicily. This study is relevant not only from a design point of view, but also from an ethical point of view, and it aims to urge engineers to show their competencies in relation to social issues.

1.5.1 Regulations and Figures on Disability

The disability issue breached the wall of indifference for the first time when the Charter of Fundamental Rights of the European Union was published in 2000 (EU 2000). In fact, Article 21 of Chapter III of the Charter prohibits any discrimination based on disability. A society that is open and accessible to all is the goal of the European Union disability strategy, for which the principle that there should be "nothing about people with disability without people with disabilities" (EORG 2004) holds. At a global level, the Convention on the Rights of People with Disabilities of United Nations (UN 2007) symbolizes the high point of governmental attention on the disability issue. The purpose of the Convention is *to promote, protect and ensure the full and equal enjoyment of all human rights and fundamental freedoms by all persons with disabilities, and to promote respect for their inherent dignity*. Moreover, this Convention formalizes the concept of "universal design" as the means to reduce the amount of adaptation of a product required to meet the specific needs of a person with disabilities to a minimum.

A census of people with disabilities is not still completely available for three main reasons. Firstly, there is no universal definition of disability, and so a unique set of indicators is not available. Since 1980, the International Classification of Impairments, Disabilities and Handicaps (ICIDH), published by the World Health Organisation (WHO), has made a distinction between impairment, disability and handicap. In 2001, the World Health Assembly adopted the International Classification of Functioning, Disability and Health (ICF). Secondly, the accuracy of the survey depends on the type of disability. Health Interview Surveys (HIS) and Disability Interview Surveys (DIS) are widely accepted instruments that can provide comparable data on health, disability and social integration. Thirdly, due to social and psychological issues, many people do not want to declare their disabilities or those of their families. Therefore, the real number of people with disabilities is probably underestimated by surveys conducted thus far.

It is currently estimated that at least 10% of the EU population will be affected at some point in their life by a disability. The Italian data are aligned with the EU data (about 15% of Italian families are involved in the disability issue) (ISTAT 2000). Surprisingly, mental health problems now account for a quarter of all disability in the EU (EU 2007).

1.5.2 Objectives of the Study and Work Plan

The objectives of this study were determined in complete accordance with doctors, paramedics and managers of OASI Maria SS. It was immediately clear that improvements in the performance, functionality and design of the postural seat system for children with mental retardation were needed. The specific needs of these patients required a postural system completely different from those of other disabled people. This system, if possible, is more difficult to settle and even more costly and ugly. Moreover, the high degree of dependence of these patients—they require constant assistance from paramedical and parents/relatives—makes easy-to-handle regulation procedures necessary.

In detail, the new postural seat system should incorporate the following features:

* Improved performance in terms of lightness and maneuverability.
* An easier manual postural regulation system.
* The presence of a diagnostic system that is able to signal departure from ideal postural settings.
* A pretty design.

The last feature was inserted to reduce the sense of "abnormality" experienced by patients and parents/relatives during the use of the postural seat system.

The study was divided into two parts. The first part, which we were responsible for, aimed at identifying the specific needs associated with the improvements and translating them into engineering suggestions for the planning phase (performed by the University of Naples "Federico II"). The methodology we followed is described in Fig. 1.7.

1.5.3 Customer Identification

The first and crucial step in achieving customer satisfaction is to clearly determine the customers and the process leading from the company to the customers (Dahlgaard et al. 1998). Nevertheless, a general definition of customers has not yet been formulated in the literature. This is due to not only the inherent differences within the goods and services sector, but also differences between studies performed within this sector. In the ISO 9000 standard, a customer is defined as an organization or a person who receives a product. This definition is restrictive for our study, because in mental or psychological disability issues the system of people around the patient play a central role in therapeutic and rehabilitative functions.

Fig. 1.7 Phases of the
methodology adopted for
the study of user needs in the
redesign of a postural seat
system

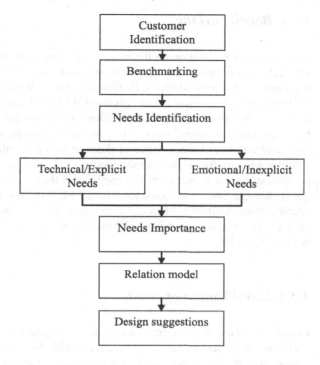

The following groups may be seen as customers for our study of the postural seat system:

- *Patients* are the real beneficiaries of the improved postural seat system. Due to their disabilities, these patients are not always able to directly express their needs.
- *Doctors* give instructions to paramedics regarding the correct postural fitting of the patients at different hours of the day.
- *Paramedics* follow the instructions of doctors regarding the regulation of the postural seat system.
- *Parents/relatives* often work as intermediaries for patient needs. Moreover, they execute the regulation of the postural seat system at home.
- *National healthcare organizations* pay for patient sanitary assistance and contribute to the purchase of the postural seat system.

Other customers include *hospitals*, where patients are cured, and *society*, which expects the new and functional postural seat system to be available on the market.

Therefore, a definition of "customer" that seems more suited to this study is *the people or the organizations that are the reason for our activities, i.e., those for whom we want to create value by our activities and products* (Bergman and Klefsjö 1994).

Because there are several potential customer categories, the various needs and expectations have to be combined through considered prioritization.

1.5.4 Benchmarking

After a careful meditation on the roles of the various people involved in this study, we made an accurate survey of the postural seat system already used by people with the same or similar disabilities. Particular attention was paid to the most brands of postural seat system most commonly used at Oasi Maria SS. The characteristics of sixteen models were examined. These characteristics were divided into eight groups: back rest, cushion, lateral push, pelvic waistband, footboard, lumbar push, armrest, and headrest. The characteristics of these groups guarantee postural functionality, stability and comfort. A frequency diagram was used to show how commonly the postural seat system characteristics appeared in the examined models. The most frequent characteristics were considered basic, while the less frequent were considered specific to or distinctive of model and brand. The use and functionality of all the characteristics were then discussed with doctors, paramedics and the technical staff at Oasi Maria SS.

1.5.5 Identification of Needs

In order to collect both explicit and technical customer needs as well as the implicit and emotional ones, we used different tools with different customers. In particular, a structured interview was used with ten doctors and paramedics because of their high degree of knowledge. In contrast, a simplified version of the Kano questionnaire, as described in Sect. 1.4, was prepared for fifteen parents/relatives of patients. Moreover, the questionnaire was integrated at the beginning with a preliminary set of questions on customers actual feelings about the postural seat system, and at the end with a set of questions on possible critical incidents (what, when, where and why they happened). The preliminary questions highlighted a poor satisfaction in the performance of the current postural seat system and the great difficulties involved in modifying the postural parameters, as suggested by doctors and instructed by paramedics.

The complete list of needs arising from doctors/paramedics and parents/relatives are reported in Table 1.1, with a distinction made between explicit and technical needs and Kansei words.

1.5.6 Importance of Needs

The importance of each identified need was calculated using the choice time methodology described in Sect. 1.4 and the JAVA interface reported in the appendix. Twenty-five respondents were subjected to the same interview. A brief introduction illustrates the aim of the survey and the general steps to follow. After providing some input data, the respondent is asked to look at the list of needs and then choose the

Table 1.1 List of customers' needs divided into explicit/technical and implicit/emotional

		Customers	Used method
	Explicit/technical needs		
1	Pathology adaptability	Doctors/paramedics	Structured interview
2	Armrest adjustability	Doctors/paramedics	Structured interview
3	Cushion anatomy	Doctors/paramedics	Structured interview
4	Body adaptability	Doctors/paramedics	Structured interview
5	Pelvis blocking	Doctors/paramedics	Structured interview
6	Reduction of the sense of weakness	Doctors/paramedics	Structured interview
7	Transportability	Parents/relative	Critical incident
8	Lightness	Parents/relative	Kano model
9	Maneuverability	Parents/relative	Critical incident/Kano model
10	Reducibility	Parents/relative	Critical incident
11	Ease of setting	Parents/relative	Kano model
	Implicit/emotional needs		
12	Comfort	Doctors/paramedics	Structured interview
13	Color & Design	Parents/relative	Kano model
14	Robustness	Doctors/paramedics	Structured interview

most preferable. A computer clock measures how long the respondent takes to make the selection. The attribute list is updated after each selection and randomized. The ranking task continues until the respondent makes the final choice between the last two needs. A common time of 1000 ms was chosen as an estimate of the reaction time t^*. The ratings session was replaced by a more simple confirmation session, in

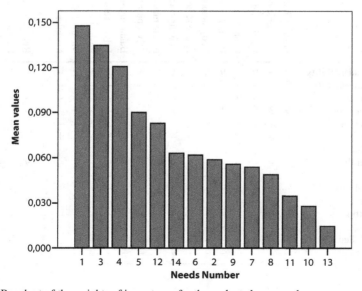

Fig. 1.8 Bar chart of the weights of importance for the evaluated user needs

which respondents were asked to look at a bar diagram representing the weights of the needs in descending order. In all cases, the respondent confirmed the results of the procedure.

The mean values of the weights of the needs are graphically represented in Fig. 1.8.

1.5.7 Relation Model

An augmented HOQ, as described in Sect. 1.4.1, was used as the model that links the customers' needs to the engineering characteristics. The engineering characteristics (technical response section) for the needs in Table 1.1 were determined with the help of medics and paramedics. In particular, the same criterion used for customer needs was followed: some engineering characteristics are related to explicit/technical needs, while others are related to implicit/emotional needs. The engineering characteristics are reported in the upper part of Table 1.2.

Table 1.2 Relationship matrix of the HOQ

	Position indicator	Electronic position system	Lumbar/ridge push system	Cushion regulation system	Balancing roll	Body support	Frame structure	Coupling with endless screw	Frame material	Pelvis push system	Balancing seat	Pelvis waistband	Push system for adduction	Mobile footboard	Modular headrest	Inclinable backrest	Releasable cushion	Deepness of seat	Washable and breathable cloth	Interchangeable cloth
Pathology adaptability	9	9	9	9	9	9	–	1	–	3	3	1	3	3	–	3	–	1	–	–
Armrest adjustability	9	–	–	–	–	–	–	3	–	–	–	–	–	–	–	–	–	–	–	–
Cushion anatomy	–	–	9	9	–	–	–	–	–	–	–	–	–	–	1	–	–	–	–	–
Body adaptability	9	3	9	–	1	9	–	1	–	–	–	–	–	–	1	–	–	9	–	–
Pelvis blocking	9	1	–	–	–	–	–	–	–	9	9	3	3	–	–	–	–	1	–	–
Reduction of sense of weakness	–	–	–	–	–	–	–	–	–	–	–	–	9	–	–	1	–	–	–	–
Transportability	–	–	–	–	–	–	9	–	–	–	–	–	–	–	–	–	–	9	–	3
Lightness	–	–	–	–	–	–	1	–	9	–	–	–	–	–	–	–	–	–	–	–
Maneuverability	–	–	–	–	–	–	9	3	–	–	–	–	–	–	–	–	–	–	–	–
Reducibility	–	–	–	–	–	–	9	–	–	–	–	–	–	–	–	–	3	–	–	–
Ease of setting	9	9	–	–	–	–	–	3	–	–	–	–	–	–	–	–	–	–	–	–
Comfort	–	–	–	–	9	–	–	–	–	–	–	–	–	–	–	–	–	–	–	–
Color & design	–	–	–	–	–	–	9	–	3	–	–	–	–	–	–	3	–	–	9	–
Robustness	–	–	–	–	–	–	9	–	–	–	–	–	–	–	–	–	–	–	–	–

The next step was to compile the planning matrix (the right part of the HOQ). The matrix we used contains the following elements:

- Importance to customer: this column records how important each need is to the customer. The data for this column are the mean values of the weights of the needs.
- Customer satisfaction performance: this is the customer's perception of how well the available postural seat system is meeting his/her needs. The values in this column were assigned by authors in accordance with doctors and paramedics (development team) on a five-point scale.
- Goal: in this column, the development team (in collaboration with experts from the University of Naples "Federico II") decided on realistic performance values for the new postural seat system. The numerical values were based on the same scale.
- Improvement ratio: this is a measure of the effort required to approach customer satisfaction performance for the defined goal. It can be calculated with several formulae (Cohen 1995), but the most simple and intuitive is the ratio of the goal to the customer satisfaction performance.
- Raw weight: this is a summary of the planning matrix. The values in this column are the product of the importance to customer column and the improvement ratio column. The higher the raw weight, the more important the corresponding customer need should be to the development team.

The columns "competitive satisfaction performance," "sales point" and "normalized raw weight" do not apply in this study and are therefore ignored here. The values of the planning matrix are reported in Table 1.3.

Table 1.3 Planning matrix for the HOQ

	Importance to customer	Customer satisfaction performance	Goal	Improvement ratio	Raw weights
Pathology adaptability	0.148	2	5	2.5	0.370
Armrest adjustability	0.121	2	2	1.0	0.121
Cushion anatomy	0.09	3	4	1.3	0.119
Body adaptability	0.015	2	5	2.5	0.037
Pelvis blocking	0.083	3	5	1.6	0.137
Reduction of sense of weakness	0.056	2	4	2.0	0.112
Transportability	0.062	2	3	1.5	0.093
Lightness	0.135	2	4	2.0	0.270
Maneuverability	0.063	2	4	2.0	0.126
Reducibility	0.035	3	3	1.0	0.035
Ease of setting	0.054	1	5	5.0	0.270
Comfort	0.059	4	4	1.0	0.059
Color & design	0.049	3	4	1.3	0.065
Robustness	0.028	1	4	4.0	0.112

The third step in the construction of the relation model is to compile the relationship section. This is a matrix where the number of rows is equal to the number of customer needs and the number of columns is equal to the number of engineering characteristics. Each cell c_{ij} contains an indication of the strength of the link between the i-th customer need and the j-th engineering characteristic. The strength of the link is usually expressed by a symbol (Asians and Americans use different symbols) and then converted into a numerical value. The numerical values were assigned according to American coding, i.e., 9 for an extremely strong relation, 3 for a moderately strong relation, 1 for a weak relation and 0 for no relation between the customer need and the engineering characteristic. The relationship matrix is reported in the central part of Table 1.2.

The last part of the relation model for this study is the row of priorities. This row summarizes the relative contributions of each engineering characteristic to the overall customer satisfaction. The priority for the j-th engineering characteristic is calculated as:

$$p_j = \sum_{i=1}^{I} c_{ij} \times r_i \tag{1.4}$$

where c_{ij} is the relation value between the i-th customer need and the j-th engineering characteristic, and r_i is the value of the raw weight for the i-th customer need.

The larger the value of the priority, the more influence the engineering characteristic has on the customer satisfaction performance, and therefore the more important it is in the development of the new model of postural seat system.

The priority values for this study are reported in Table 1.4. The last three sections of the HOQ, i.e., competitive benchmarking, targets and technical correlation, apply in a subsequent phase of product development process, when the created product concept is evaluated in comparison with those of its competitors, or when production constraints force designers and engineers to solve the potential correlations between technical characteristics of the product.

1.5.8 Design Suggestions

The process followed allowed a list of design interventions to be defined based on customer needs. Interpreting the results summarized in Table 1.4, it was possible to suggest the strategic elements needed to improve the development of an innovative postural seat system to the designers and engineers at the University of Naples. These elements are (in order):

- Indicators: they should be easy to see and handle (characteristic 1).
- Postural regulation systems: they should be automatic, to facilitate parents/tutors in setting them (characteristics 2-3-4-5-6-8-13-14).
- Structure: new materials (light and robust) should be used, and a new structure (reducible and transportable) should be developed (characteristics 7 and 9).

Table 1.4 Priorities for engineering characteristics

	Engineering characteristics	Priorities
1	Position indicator	8.42
2	Electronic system for position	6.01
3	Lumbar/ridge push system	4.73
4	Cushion regulation system	4.40
5	Balancing roll	3.89
6	Body support	3.66
7	Frame structure	3.14
8	Coupling with endless screw	2.97
9	Frame material	2.63
10	Pelvis push system	2.34
11	Balancing seat	2.34
12	Pelvis waistband	1.78
13	Lateral push system for adduction	1.52
14	Inclinable backrest	1.30
15	Mobile footboard	1.27
16	Modular headrest	1.12
17	Releasable cushion	0.94
18	Deepness of seat	0.84
19	Washable and breathable cloth	0.59
20	Interchangeable cloth	0.28

- Seat: this is one of the most important parts of the postural seat system. The characteristics 10, 11, 12 and 18 should be significantly improved.
- Adaptability: characteristics 15 and 16 indicate the need to develop a modular headrest and a free-to-move footboard.
- Versatility: new fabrics and interchangeable parts could improve the design of the postural seat system and its versatility in use.

1.6 Conclusions

The most important step in the design process is the initial one, in which customer needs are identified and examined. The need for manufacturers to capture and correctly interpret the requirements of their target customers has led to the development of a number of techniques aimed at bringing the "voice of the customer" into the design process. The voice of the customer, the Kano model and quality function deployment have been broadly used in the design and development product process. Nevertheless, a successful product development process should be based not only on the stated/explicit customer needs, but also on the implicit needs and feelings of the customer. Kansei engineering is a methodology that makes it possible to incorporate customers' emotions and perceptions into the product design process.

The integration of QFD and Kansei engineering methodology can lead to an increased customer satisfaction, since the product should fulfill both the expected and the emotional needs of the customer. Moreover, there are real situations in which a simplified version of the theoretical tools or the creation of new practical ones is strongly suggested.

In this work, we propose a general framework for capturing customers' needs for the design of an innovative postural seat system for patients affected by mental retardation. A step-by-step procedure, carried out in collaboration with doctors, paramedics, managers, technicians and parents/tutors of Oasi Maria SS, allowed strategic elements of the design intervention to be defined. These elements will be improved by the design team of the University of Naples.

Acknowledgements The authors thank the engineers Virna Lo Monaco and Sebastiano Ferrigno for their technical support throughout the process of needs analysis and are grateful to Dr. Arturo Caranna, head of the Quality Department of Oasi Maria SS, for his invaluable support. Special thanks are directed to all of the people working at Oasi Maria SS, who gave us helpful comments and support.

References

Akao, Y.: New product development and quality assurance—quality deployment system. Stand. Qual. Control **25(4)**, 7–14 (1972)

Akao, Y.: Quality Function Deployment, Integrating Customer Requirements into Product Design. Productivity Press, Cambridge (1990)

Akao, Y., Mazur, G.H.: The leading edge in QFD: past, present and future. Int. J. Qual. Reliab. Manag. **20(1)**, 20–35 (2003)

Arnold, K.: Kansei Engineering and Quality Function Deployment: Towards Increased Customer Satisfaction (LITH-IKP-EX-1868). Linköping Univ., Linköping (2000)

Barone, S., Lombardo, A.: Service Quality Design Through a Smart Use of Conjoint Analysis. Asian J. Qual. **5(1)**, 34–42 (2004)

Barone, S., Lombardo, A., Tarantino, P.: A Weighted Logistic Regression for Conjoint Analysis and Kansei Engineering. Qual. Reliab. Eng. Int. **23**, 689–706 (2007)

Bergam, B., Klefsjö, B.: Quality. Studentlitteratur, Lund (1994)

Center for Quality Management: Kano's methods for understanding customer-defined quality. Center Qual. Manag. J. **2(4)**, 3–36 (1993)

Chan, L.C., Wu, M.L.: A systematic approach to quality function deployment with a full illustrative example. Omega **33(2)**, 119–139 (2005)

Cohen, L.: Quality Function Deployment—How to Make QFD Work for You. Addison-Wesley, Reading (1995)

Crosby, P.B.: Quality is Free: The Art of Making Quality Certain. New American Library, New York (1979)

Dahlgaard, J.J., Kristensen, K., Kanji, G.K.: Fundamentals of Total Quality Management. Nelson Thornes Ltd., Cheltenham (1998)

Deming, W.E.: Out of Crisis. Massachusetts Institute of Technology, Cambridge (1986)

European Opinion Research Group EEIG: The European Year of People with Disabilities 2003. Special Eurobarometer 198 (2004)

Flanagan, J.C.: The critical incident technique. Psychol. Bull. **51(4)**, 327–358 (1954)

Fonti, V., Tarantino, P.: Statistical Methods for Kansei Engineering: Application to Product and Service Design. In: Proceedings of ADM Conference "Conceptual Design Methods for Product Innovation," Forlì, Italy, 14–15 Sept (2006)

Govers, C.P.M.: What and how about quality function deployment (QFD). Int. J. Prod. Econ. **46–47**, 575–585 (1996)

Gremler, D.D.: The critical incident technique in service research. J. Service Res. **7(1)**, 65–89 (2004)

Griffin, A., Hauser, J.R.: The voice of the customer. Market. Sci. **12(1)**, 1–23 (1993)

Gustafsson, A., Herrmann, A., Huber, F.: Conjoint Measurement—Methods and Applications, 3rd edn. Springer, Berlin (2003)

Hauser, J.R., Clausing, D.: The house of quality. Harvard Bus. Rev. **December**, 63–73 (1988)

Hinterhuber, H.H., Matzler, K.: How to make product development projects more successful by integrating Kano's model of customer satisfaction into quality function deployment. Technovation **18(1)**, 25–38 (1998)

ISTAT: Indagine sulle condizioni di salute e ricorso ai servizi sanitari. ISTAT, Rome (2004)

Jaccard, J., Brinberg, D., Ackerman, L.J.: Assessing attribute importance: a comparison of six methods. J. Consum. Res. **12 (March)**, 463–468 (1986)

Kano, N., Seraku, F., Takanashi, F., Tsuji, S.: Attractive quality and must-be quality. J. Jpn. Soc. Qual. Control **14(2)**, 39–48 (1984)

Keniston, K.: The crisis of the engineering algorithm. Talk for the Institute of Advanced Studies in Humanities, Politecnico di Torino, 17 Oct. (1996)

Khalid, H.M., Helander, M.G.: A framework for affective customer needs in product design. Theor. Issues Ergonom. Sci. **5(1)**, 27–42 (2004)

Lanzotti, A., Tarantino, P.: Kansei Engineering Approach for Identifying Total Quality Elements. In: Proc. 10th QMOD Conf., Helsingborg, Sweden, 18–20 June (2007)

Lee, S., Harada, A., Stappers, P.J.: Pleasure with products: design-based Kansei. In: Green, W., Jordan, P. (eds) Pleasure with Products: Beyond Usability, pp. 219–229. Taylor & Francis, London (2002)

Lockamy, A., Khurana, A.: Quality function deployment: total quality management for new product design. Int. J. Qual. Reliab. Manag. **12(6)**, 73–84 (1995)

Löfgren, M., Wittel, L.: Kano's theory of attractive quality and packaging. Qual. Manag. J. **12(3)**, 7–20 (2005)

Mazur, G.H.: Voice of customers analysis: a modern system of front-end QFD tools. Proceedings of the 10th ASQ Congress. American Society of Quality Control, Milwaukee, WI (1997)

Nagamachi, M.: Kansei engineering: a new ergonomic consumer-oriented technology for product development. Int. J. Ind. Ergonom. **15**, 3–11 (1995)

Nagamachi, M., Matsubara, Y.: Hybrid Kansei engineering system and design support. Int. J. Ind. Ergonom. **19**, 81–92 (1997)

Norman, D.A.: Emotional Design: Why We Love (or Hate) Everyday Things. Basic Books, New York (2004)

European Union: Character of Fundamental Rights of the European Union. Offic. J. Eur. Union **43**, 18 December 2000

Park, J., Han, S.H.: A fuzzy rule-based approach to modeling affective user satisfaction towards office chair design. Int. J. Ind. Ergonom. **34**, 31–47 (2004)

Schütte, S.: Engineering Emotional Values in Product Design, Kansei Engineering in Development (Linköping Studies in Science and Technology, Dissertation No. 951). Linköping Univ., Linköping (2005)

Schütte, S., Eklund, J.: Design of rocker switches for work-vehicles: an application of Kansei engineering. Appl. Ergonom. **36**, 557–567 (2005)

Sullivan, L.P.: Quality function deployment—a system to assure that customer needs drive the product design and production process. Qual. Progr. **June**, 39–50 (1986)

Tague, N.R.: The Quality Toolbox, 2nd edn. ASQ Quality Press, Milwaukee, WI (2004)

Toft, Y., Howard, P., Jorgensen, D.: Human-centred engineers—a model for holistic interdisciplinary communication and professional practice. Int. J. Ind. Ergonom. **31**, 195–202 (2003)

United Nations: Convention on the Rights of Person with Disabilities and Optional Protocol, 30 March 2007. UN, New York (2007)

Chapter 2
Statistical Design for Innovation in Virtual Reality

Antonio Lanzotti, Giovanna Matrone, Pietro Tarantino,
and Amalia Vanacore

Abstract In this chapter, an original design strategy for product innovation is presented. This strategy is based on a continuous innovation process and takes advantages of both emotional design methodologies and participative design tools in virtual reality (VR). It combines techniques for user need identification and virtual reality experiments to simulate user-product interaction. This original combination of techniques allows the early evaluation of the quality of new product concepts, which is essential for the success of innovation.

To show the main phases of this strategy, three case studies are briefly introduced. In the first case study, concerning the design of a railway coach interior, the importances of both the identification of user needs and an evaluation session in VR are highlighted. In the second case study, concerning the railway seat design, the techniques used to generate new concepts and to choose the optimal one among them are briefly illustrated. Finally, in the third case study, concerning the innovation of

Antonio Lanzotti
Department of Aerospace Engineering, University of Naples Federico II
P. le Tecchio 80, 80125 Naples, Italy
e-mail: antonio.lanzotti@unina.it

Giovanna Matrone
Department of Aerospace Engineering, University of Naples Federico II
P. le Tecchio 80, 80125 Naples, Italy
e-mail: giovanna.matrone@unina.it

Pietro Tarantino
Department of Aerospace Engineering, University of Naples Federico II
P. le Tecchio 80, 80125 Naples, Italy
e-mail: pietro.tarantino@unina.it

Amalia Vanacore
Department of Aerospace Engineering, University of Naples Federico II
P. le Tecchio 80, 80125 Naples, Italy
e-mail: amalia.vanacore@unina.it

Pasquale Erto, *Statistics for Innovation*
ISBN 978-88-470-0814-4, © Springer 2009

a postural seat system, the continuous innovation iterative cycle—which is crucial
to the statistical design for innovation strategy—is described, starting from designer
sketches.

2.1 Introduction

Recent tendencies in the field of product design show increasing interest in objects
that are able to inspire users and to evoke positive feelings and emotions (Demirbilek
and Sener 2003). The designer's creativity alone cannot provide sufficient support
for the complex process of understanding the emotions arising from the user–object
interaction and coding them as useful information for successful design. The need
for a new design strategy is particularly evident in the concept design phase where,
to a great extent, the future success of a product is determined. In this phase, the
designers look for a concise description of all product features that will satisfy user
needs (Ulrich and Eppinger 2000). Consequently, the early correct identification of
these needs plays a central role in the maximization of user satisfaction (Di Gironimo et al. 2006). In the last few years, several methodologies have been developed
to identify the functional user needs as well as the emotional ones. Some of these
methodologies apply a user-centered approach, a designer-driven approach, or a balanced mix of the previous ones, in order to satisfy conscious and unconscious user
needs and translate them into the functional features of the product. Among them,
the Kano model and Kansei engineering are widely applied in research as well as industry. The Kano model allows designers to classify product functional elements in
quality categories (Kano et al. 1984). Kansei engineering, being an emotional design
methodology (Norman 2004), allows designers to objectively understand the human
subjective and psychological sensibility and connect it to the process of product development (Nagamachi 1995). Once user needs are identified, designers develop
new product concepts in compliance with standard engineering constraints. At this
stage, the combined use of computer-aided styling (CAS) and computer-aided design (CAD) tools, experimental statistical methods and a top-down approach (Di
Gironimo et al. 2008) to explore concept alternatives by means of virtual prototypes
is essential. The generated alternatives are tested and the best solution is identified.
In this phase, the opportunities offered by virtual reality can overcome the main limitations of traditional approaches that rely on physical prototypes (Ottosson 2002).
In particular, the availability of a virtual prototype allows designers to anticipate
aesthetic, ergonomic and usability tests in the concept design phase by means of
a participatory design and, at the same time, to reduce the development time and
costs of physical prototypes (Bruno and Muzzupappa 2006).

In this chapter, a systematic design strategy for improving the quality of new designed products is presented. Section 2.2 illustrates the logical phases of the new
strategy, which takes advantage of emotional design methodologies as well as participative design in virtual reality. Section 2.3 briefly shows the first results from the
application of the proposed strategy in three significant case studies.

2.2 From Emotions to Innovation in Virtual Reality

Innovation is realized through the introduction of product features that address new user needs or provide new solutions in order to satisfy—significantly more so than previously—pre-existing user needs. The proposed strategy, denoted "statistical design for innovation," is based on the continuous identification of both conscious (i.e., expressed) and unconscious (i.e., latent) user needs and their translation into product features called quality elements (QEs). The translation of user needs into QEs is one of the most critical phases in new product design. Up to now, this translation was achieved through either designer creativity or methodologies based on expert user involvement. The statistical design for innovation strategy recognizes the importance of both approaches and combines them in order to identify the best product architecture. It integrates a recently proposed concept design for quality procedure (Di Gironimo et al. 2006) with the emotional design approach (Nagamachi 1995), (Lanzotti and Tarantino 2007). The logical phases of the proposed strategy are depicted in Fig. 2.1.

The proposed strategy is realized through five phases, which are briefly illustrated in the following.

I.a   *Identification of user needs.* The product design process begins with the identification of both expressed and latent user needs. This phase is critical because of the influence of the emotional user process on the success of the product. The Kano model and Kansei engineering (KE) are the methodologies applied here to accomplish the task of identifying user needs. In particular, the Kano model allows us to identify quality elements that satisfy the expressed user needs. It starts with a small group (10–15) of QEs proposed by designers and then assigns priorities to 3–5 elements by classifying these QEs into quality categories (Kano et al. 1984). On the other hand, KE allows us to

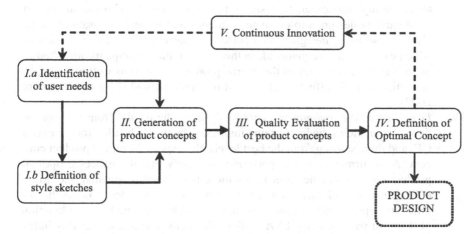

Fig. 2.1 Logical phases of the statistical design for innovation in VR strategy

identify the latent and emotional quality elements and to translate them into physical solutions for product concept (Nagamachi 1995). KE starts with the identification of an almost exhaustive group (70–80) of QEs and then allows us to select the most meaningful (3–5) among them according to the feelings and emotions of a group of potential users, by means of Pareto analysis, factor analysis and ordinal logistic regression.

I.*b*   *Definition of style sketches.* The designer takes part in the statistical design for innovation process by developing style and product architecture ideas starting from the previously identified QEs and by using simple tools such as pencils and colors. In this phase, the designer can ascertain some needs not identified in the previous phase and also interpret the user needs so as to add originality to the QEs definition. This process requires a knowledge of the market, patents, materials and new trends in the fields of art and style, as well as a knowledge of standard constraints, which are very important for industrial products.

II.   *Generation of product concepts.* Starting from the identified QEs and style sketches proposed by designers, product concepts can be generated through the use of DOE statistical tools. The main constraint is the maximum number of virtual prototypes that can be evaluated by expert users in a participative design session. For this reason, fractionated factorial designs are used and only a few levels (2–3) are adopted for each QE (Wu and Hamada 2000). Each QE level is developed from sketches into a CAS–CAD (computer-aided styling–computer-aided design) model following the style architecture eventually defined by designers. Thus, the output of this phase is a limited number (4–8) of virtual prototypes of product concepts developed by following the adopted fractionated factorial design.

III.   *Quality evaluation of product concepts.* In this phase, the generated virtual prototypes are transferred to the virtual reality environment in order to allow expert user evaluation. The VR tools allow the user–product interaction to be tested through photorealistic visualization in an immersive environment that is adequately designed to evaluate aesthetics and to perform usability tests. Using these tools, designers can perform a participatory design session in which expert users are involved. In this session, the participants are asked to express their opinions about the overall product concepts as well as each QE, according to main effects analysis and the EVA method (Erto and Vanacore 2002).

IV.   *Definition of the optimal concept.* Analysis of the results from the evaluation session allows the level of maximum satisfaction to be identified for each QE, and then, starting from the best levels, the expected optimal product concept. A confirmatory test is performed to verify that the expected optimal product concept satisfies user needs more than the experimental one. Main effects analysis and the EVA method can be adopted in order to assess the optimal concept, identifying the most important QEs and the best combination of them. In particular, the EVA method allows us to evaluate a quality index that reflects the level of user satisfaction for each significant QE.

Table 2.1 Statistical techniques and design tools involved in each phase of the proposed product innovation strategy

| | Phases | | | |
	I	II	III	IV
Statistical techniques	– Pareto diagram – Factor analysis – Ordinal logistic regression	– DOE	– Main effects analysis – EVA method	– Main effects analysis – EVA method
Design tools	– Creativity – Pencil	– Pencil – CAS–CAD – virtual prototypes	– VRLab – Interaction with Virtual prototypes	– VRLab

V. *Continuous innovation.* At the end of phase IV, designers can verify whether innovation really has been achieved. If it has, the design strategy suggests that another iteration of the procedure should be performed in order to satisfy other needs, not contemplated in the first experimentation. In contrast, if innovation has not been achieved, the procedure is reactivated and designers look for the reasons for the lack of success, which can be either incomplete identification of the user needs or—more often—the wrong translation of user needs into product concepts. This process, taken together, allows to improve the success of innovative products.

The statistical design for innovation strategy involves the application of statistical methodologies to help designers synthesize the input from expert users into new features of virtual prototypes. Table 2.1 shows the statistical techniques and design tools involved in each phase of this product innovation process. Some phases are particularly important to the innovation process: the identification of user needs could be incomplete, ignoring important needs, or designers may have chosen to study first some needs and then other ones. Moreover, in the generation phase, the levels chosen for each QE or the product architecture selected could be wrong. For these reasons, at the end of the experimentation stage, a continuous innovation process is activated. Based on the collected information, the collaborative process between designers and expert users is updated until a sufficiently improved result is obtained.

The innovation process is achieved through the use of virtual and physical prototypes used in product and process simulation. Robust design techniques are proposed in order to complete the product development with the implementation of physical solutions.

2.3 Three Applications of Statistical Design for Innovation

The proposed design strategy and its operational phases are now presented by describing three case studies in the next three sections. Each case study is briefly introduced to illustrate, following an applicative approach, some particular phases of the new strategy.

2.3.1 Railway Coach Interior Design

This case study was developed to improve the interior design of a railway coach. It originated as a fruitful collaboration between the Department of Aerospace Engineering, the Department of Engineering Design and Industrial Management of the University of Naples Federico II, Firema Trasporti S.p.A., and the Competence Center for the Qualification of Transportation Systems (founded by the Campania Region), which hosts a virtual reality laboratory called VRTest. Only the most innovative phases of the case study are discussed here: the identification of user needs and concept evaluation in VR. Further details about the case study can be found in Lanzotti and Tarantino (2007) and Di Gironimo et al. (2007).

2.3.1.1 Identification of User Needs

A simplified version of Kansei engineering was used to identify the quality elements associated with the emotional perspective of users. Passengers of a medium-haul train were chosen as the target group of this study. By scanning several sources of information, thirty-nine words describing the emotional bond between passengers and train interiors were identified (Kansei words). These words were reduced to a more manageable number using both factor analysis and affinity diagrams. The final set of words was: *comfort, originality, mobility, versatility, simplicity*. A similar procedure was used to identify the functional requirements to be studied: *closed-circuit monitoring system, recyclability, support for standing passengers, wide spaces, windows*. Each of these requirements was represented by two alternatives (i.e., levels) with respect to Italian railway standards. These levels were arranged in a factorial design to estimate the correlation between functional requirements and Kansei words. Starting from the mood boards technique (McDonagh et al. 2002), an innovative procedure to exhibit concepts in KE sessions was used. According to this new procedure, sketches representing different combinations of functional requirement levels were created using minimum design elements. These sketches aimed to hint at the design of the railway coach interior rather than to show all of its properties. Thirty students enrolled at the Faculty of Industrial Design of the Second University of Naples were asked to rate their impressions of each product concept on a five-grade Likert scale. It was pointed out that the first impression was the most important. No time constraint was considered.

In order to analyze the collected data, ordinal logistic regression (OLR) was applied. This method, a general linear model, has been proposed for use in the case of Likert scale data (Lawson and Montgomery 2006). The results from OLR indicated that the Kansei word *comfort* was more associated with the QE (quality element) *support for standing passengers* than the elements *closed-circuit monitoring system* and *wide spaces*, whereas the Kansei word *mobility* was moderately associated with the element *wide spaces*. No other association between Kansei words and functional requirements was found to be statistically significant.

2.3.1.2 Quality Evaluation in VR

Support for standing passengers was the QE that was most strongly connected to the sense of *comfort*. Following usability principles, three particular design factors were chosen for this QE: shape, position and color. In particular, these factors were studied for hands, perches and handrails. For each factor, three design solutions were designed. Nine concepts were first drawn in CAD according to a nine-run fractional factorial design, which were subsequently transferred into virtual reality software. The generated train interior concepts were improved by applying light and texture in order to increase their realism, before they were evaluated in an immersive virtual reality environment. Ten expert users took part in the evaluation session. They were first asked to express the feelings of *comfort* evoked by the different design solutions on a ten-grade Likert scale, and then to answer a series of questions in order to implement the EVA method (Erto and Vanacore 2002). This method quantitatively measures the specific contribution of each physical quality element to the global quality level of a chosen product concept. Figure 2.2 shows a moment during the evaluation phase at the VRTest laboratory.

A confirmatory session was performed, within one week, in order to verify the robustness of the procedure and the coherence of the results. On the basis of the expert user responses in both sessions (planned and confirmatory), the optimal concept shown in Fig. 2.3 was identified. It yielded a 40% improvement in perceived quality compared with the concept designed without using the proposed strategy.

In conclusion, by using the proposed strategy, it was possible to identify a new design element that was able to improve the users' sense of *comfort* as well as the global quality perception of train interiors.

2.3.2 Railway Seat Design

This case study concerns the seat design for a regional train; it was developed at the virtual reality laboratory of the Competence Center for the Qualification of Transportation Systems founded by the Campania Region. It shows how the proposed approach allows the deep separation between the activities of styling and engineering to be reduced. The design cycle starts with both the sketches proposed by design-

Fig. 2.2 The expert users involved in the evaluation session in the VRTest laboratory

Fig. 2.3 The optimal concept identified using statistical design for innovation strategy

ers and the identification of user needs; it proceeds with assembly modeling and the generation of several virtual prototypes of railway seats in compliance with the constraints of standards; and it ends with the identification of the best concept for the railway seats obtained through evaluations made by expert users in the virtual reality environment. This procedure allows a realistic and reliable verification of the results due to the expert users' involvement in the main phases of product development. In the following, we will focus on the generation of product concepts and the definition of the optimal one. A more detailed description of the case study can be found in Di Gironimo et al. (2008).

2.3.2.1 Generation of Product Concepts

In the first phase, three different style solutions, named *yesterday*, *today* and *tomorrow*, were developed for new railway seats. Four QEs were identified from a Kano analysis (*armrests, direct lighting, footrest, tip-top table*) and, for each QE, two alternative solutions were defined according to railway standards. In order to test the identified QE for each style, and taking into account the maximum number of virtual prototypes that can be evaluated by expert users in a participative design session, a 2^{4-1} fractional factorial design was adopted (Table 2.2).

Afterwards, the twenty-four seats concepts were modeled in a 3D CAD environment. CAD software and the top-down approach were used to develop the virtual prototypes according to a datum structure for assemblies and parts (Whitney 2004). Some suitable geometric references (points, axes, planes) were built, based on the standards. These references allowed the features of the QE to be realized and subsequently modified in an easy and continuous updating process. The study of the standards was the first step in implementing the datum structure; the geometric references were used to control the main parameters of seats such as height, width and depth. Due to the datum structure, the generated concepts complied with both the standards and the first style ideas. In order to guarantee a good immersive experience in the evaluation phase, a basic seat was designed for each style and then the QEs were added according to the respective level in the chosen design (Fig. 2.4).

Table 2.2 The experimental design (2^{4-1})

Concept	A: Armrests	B: Lighting	C: Footrest	D: Table
1	Fixed	Rear	No	Yes
2	Fixed	Rear	Yes	No
3	Fixed	Lateral	No	No
4	Fixed	Lateral	Yes	Yes
5	Moving	Rear	No	No
6	Moving	Rear	Yes	Yes
7	Moving	Lateral	No	Yes
8	Moving	Lateral	Yes	No

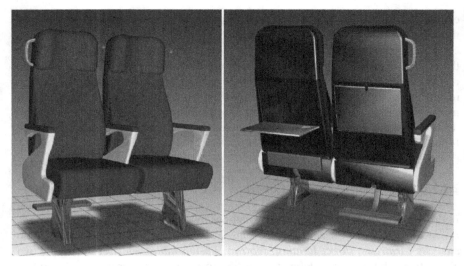

Fig. 2.4 Concept 4 for the style *today* (*armrests*: fixed; *lighting*: lateral; *footrest*: yes; *table*: yes)

2.3.2.2 Definition of the Optimal Concept

The data from the evaluation session were analyzed via Pareto ANOVA (Park 1996) and mean effects analysis. The *footrest* and the *tip-top table* proved to be the most important QEs; indeed, whatever the style, they provided the maximum contribution to the perceived quality level. Starting from the best level of each QE, the expected optimal seats concept was identified for each style. Finally, a confirmatory test was performed for each style in order to assess the expected results under homogeneous conditions. A comparison between the data from the first and second evaluation sessions indicated that *yesterday* was the best style; its optimal concept is shown in Fig. 2.5.

The EVA method was applied in order to establish the effective quality improvement in the identified optimal concept. The results of the analysis were not satisfactory for the must-be element *direct lighting*; for this reason new solutions and further experimentation were needed.

2.3.3 Postural Seat System Innovation

This case study concerns the innovation of a postural seat system for disabled children suffering from infantile cerebral paresis. It was developed in cooperation with the IRCSS Associazione Oasi Maria SS. in Troina (EN, Sicily), Italy, which is an Italian Institute involved in genetic research on mental disease recognized by World Health Organization.

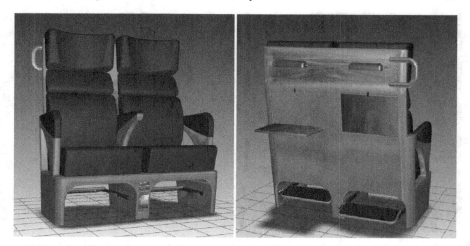

Fig. 2.5 The optimal concept for *yesterday*, identified as the best style

This case study shows that continuous innovation is marked by the continuous involvement of expert users at each stage of design in order to achieve an optimal product in terms of user satisfaction. In the following, the focus is on describing the sketch definition phase and on the continuous innovation approach.

2.3.3.1 Definition of Sketches

The designers worked intensely on style definition because of the strong influence of aesthetics on user quality satisfaction. The aim was to free the children from their "heaviness" resulting from the handicap and to spread an idea of mobility and agility. The inspiration for this came from nature: a butterfly. Many sketches were drawn for covers and control systems based on this idea (Fig. 2.6). The shape of a butterfly wing was used in sketches of the line of the back of the seat; flowers (on which butterflies rest) suggested a form for seat knobs; and butterfly feelers provided the inspiration for collocating two webcams for video surveillance.

2.3.3.2 Continuous Innovation

In this case study, the continuous involvement of expert users was essential in order to achieve successful innovation. Indeed, only those people who face the disease every day can provide an effective contribution to product innovation. For this reason, the designers chose to advance step-by-step, alternating the development of new quality features with expert user evaluation. In particular, concept generation stages and participative design sessions were iterated until a product concept that

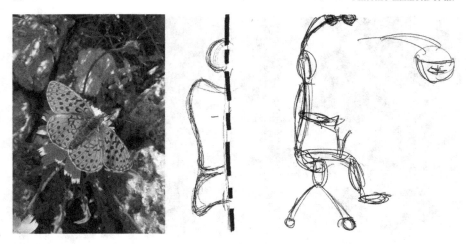

Fig. 2.6 From the concept of a butterfly to the first sketches by designer Francesco Fittipaldi (photo: Costantino Tedeschi, WWF Sannio, Italy)

was more satisfying to users than previous implementations, and with innovative features, was achieved. This design strategy allowed design improvements at every iteration of the procedure (Fig. 2.7).

The identification of user needs was critical because, in this case, the users could not express opinions on their own. For this reason, the designers focused their attention on the users' care-givers: parents and doctors. In the first phase, a lot of interviews were conducted in the Oasi in order to identify the deficiencies of pos-

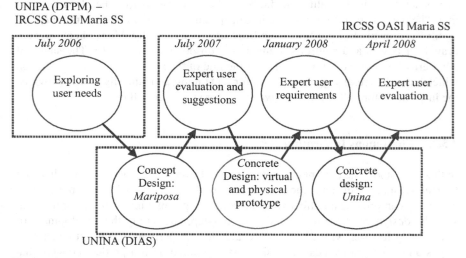

Fig. 2.7 The continuous innovation cycle used in the postural seat system statistical design

Fig. 2.8 *Mariposa* prototype

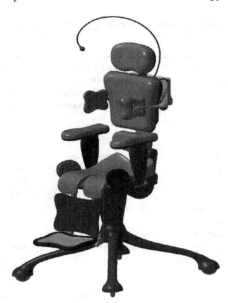

tural systems that were used in the institute. Starting from the best product in class (*BIC*) among those used in the Oasi, new concepts were generated, with several solutions for the mechanics and aesthetics. A lot of alternative concepts were studied; the most functional ones among these were selected.

An evaluation session was performed in the Oasi in order to validate the design and identify the best solution. The results showed a high satisfaction with the new solution, called *Mariposa* (Fig. 2.8), compared with *BIC*.

Then, starting from *Mariposa*, a new virtual prototype was designed to concretely define the parts and the assembly in terms of shape, dimensions and materials. This virtual prototype concept was evaluated in the Oasi in order to discover the functional requirements for innovative adjustments. Then a physical prototype was realized to test the characteristics of some adjustments that are impossible to simulate in VR. Based on the evaluation session and the analysis of the physical prototype, a new virtual prototype called *Unina* was concretely designed, incorporating all of the technical details for each part of the assembly. A final evaluation session was performed in the Oasi; the results showed an effective improvement in the proposed postural system. In particular, the *Unina* prototype nearly doubled the user satisfaction compared with the *BIC* model. Table 2.3 lists some details about the product feature innovations that resulted in this marked improvement in user satisfaction for the Mariposa and *Unina* prototypes over the the *BIC* model.

Table 2.3 Improvement of user satisfaction via product features innovation

	BIC	Mariposa	Unina
Modular systems	– Separate systems for body parts – Scarce segmentation – No balancing system	– Partial segmentation – No balancing system – 5 discrete adjustments – 2 continuous adjustments	– High segmentation – Balancing system – Adjustments for ischiatic tuberosities – 4 discrete adjustments – 5 continuous adjustments
Adjustment interface	– Adjustment levers – No position markers	– Control panel in the back – Position markers – 10 knobs	– Control panel in the back – Position markers – 1 handle
Stability of the adjustments	– Scarce	– Medium	– High
Ease of adjustment	– No	– Control panel in the back – Position markers – 10 knobs	– Control panel in the back – Position markers – 1 handle
Safety	– Presence of dangers for doctors, parents and patients	– Absence of dangers for doctors, parents and patients	– Absence of dangers for doctors, parents and patients
Aesthetics	– No	– Yes	– Yes

2.4 Conclusions

In this chapter, a design strategy based on the integration of the traditional procedure for user-centered design with the new principles of emotional design was presented. Statistical methods and quality evaluation tools can support designers in attempts to systematize the process of translating expressed/technical needs and latent/emotional needs into product features, here called "quality elements." Moreover, these methods provide designers with objective information that can be used to continuously update the design procedure and then to improve the quality of the designed products and the overall level of user satisfaction.

This strategy makes use of statistical methods, and tests the generated concept alternatives in a virtual reality environment. Extensive use of virtual reality gives designers the advantage of being able to improve quality early on in product development. The case studies show how the proposed strategy can support and integrate the designer's creativity in the innovation process, enhancing the chances of success of new products, since it increases user satisfaction.

Acknowledgements The authors thank the engineers Francesca Galileo, Vincenzo Parlato, Alejandro Cadarso Varela and Simone Pontillo for their technical support, the designer Francesco Fittipaldi for his contribution to sketch definition, Prof. Eng. Giuseppe Di Gironimo, responsible for the VRTest Lab, Dr. Arturo Caranna, and the medical staff at IRCSS OASI Maria SS. for their precious collaboration.

References

Bruno, F., Muzzupappa, M.: Participative design and virtual reality. In: Proc. Giornate di Studio ADM, Forlì, Italy, 14–15 Sept (2006)

Erto, P., Vanacore, A.: A probabilistic approach to measure hotel service quality. Total Qual. Manag. **13(2)**, 165–174 (2002)

Demirbilek, O., Sener, B.: Product design, semantics and emotional response. Ergonomics **46(13–14)**, 1346–1360 (2003)

Di Gironimo, G., Lanzotti, A., Vanacore, A.: Concept design for quality in virtual environment. Comput. Graph. **30**, 1011–1019 (2006)

Di Gironimo, G., Guida, M., Lanzotti, A., Vanacore, A.: Concept design for quality in VR: case study on the interiors of a railway coach. In: Proc. 6th IPMM Conf., Salerno, Italy, 25–29 June (2007)

Di Gironimo, G., Lanzotti, A., Matrone, G., Papa, S., Tarantino, P.: Concept design of a railway coach: quality elements identification through a Kansei-VR approach. In: Proc. 6th IPMM Conf., Salerno, Italy 25–29 June (2007)

Di Gironimo, G., Lanzotti, A., Matrone, G., Patalano, S., Renno, F.: Modellazione di assiemi ed esperimenti in Realtà Virtuale per migliorare la qualità di nuovi sedili ferroviari. Proc. 20th INGEGRAF Conf., Valencia, Spain, 4–6 June (2008)

Kano, N. et al.: Attractive quality and must-be quality. J. Jpn. Soc. Qual. Control **14(2)**, 39–48 (1984)

Lanzotti, A., Tarantino, P.: Kansei engineering approach for identifying "total quality" elements. Proc. 10th QMOD Conf. Helsingborg, Sweden, 18–20 June (2007)

Lawson, C., Montgomery, D.C.: Logistic regression analysis of user satisfaction data. Qual. Reliab. Eng. Int. **22**, 971–984 (2006)

McDonagh, D., Bruseberg, A., Haslam, C.: Visual product evaluation: exploring user's emotional relationship with products. Appl. Ergonom. **33**, 231–240 (2002)

Nagamachi, M.: Kansei engineering: a new ergonomic consumer-oriented technology for product development. Int. J. Ind. Ergonom. **15**, 3–11 (1995)

Norman, D.A.: Emotional Design: Why We Love (or Hate) Everyday Things. Basic Books, New York (2004)

Ottosson, S.: Virtual reality in the product development process. J. Eng. Des. **13(2)**, 159–172 (2002)

Park S.H.: Robust Design and Analysis for Quality Engineering. Chapman & Hall, London (1996)

Ulrich, K.T., Eppinger, S.D.: Product Design and Development. McGraw-Hill, New York (2000)

Whitney, D.E.: Mechanical Assemblies. Oxford University Press, New York (2004)

Wu, C.F.J., Hamada, M.: Experiments—Planning, Analysis and Parameter Design Optimization. Wiley, New York (2000)

Chapter 3
Robust Ergonomic Virtual Design

Stefano Barone and Antonio Lanzotti

Abstract From the early development phases of a new industrial product, realistic simulations can be performed in a virtual environment to study the human–machine interaction. In a virtual lab, it is possible to perform experiments to assess the ergonomics of the new product using mannequins simulating the human body, and to deal with the problem of anthropometric variation.

Although such sophisticated tools are available, there is still the need for a methodological framework that efficiently organizes the experiments in the virtual lab.

This paper provides an overview of robust ergonomic virtual design (REVD), a methodology developed by the authors over the last few years. It allows products to be developed with ergonomic performances that are as insensitive as possible to anthropometric variations during their life cycles. This methodology focuses on finding the optimal values for the main design parameters and, when necessary, the adjustment parameters. Furthermore, a new way of dealing with anthropometric noise is proposed. Some applications are presented, along with detailed analysis.

3.1 Introduction

Designing the ergonomics of a new product is an increasingly important aspect of early product development phases. This is particularly true if the end-user must interact with the product for long periods (Dainoff 2003). Today, the main fields

Stefano Barone
Department of Technology, Production and Managerial Engineering
University of Palermo, Viale delle Scienze, 90128 Palermo
e-mail: stefano.barone@unipa.it

Antonio Lanzotti
Department of Aerospace Engineering, University of Naples Federico II
P. le Tecchio 80, 80125 Naples, Italy
e-mail: antonio.lanzotti@unina.it

Pasquale Erto, *Statistics for Innovation*
ISBN 978-88-470-0814-4, © Springer 2009

that are developing new tools for the study of the human–machine interface (HMI) are:

- transportation (improving the comfort of cars, trains, airplanes, etc.)
- furniture (reducing discomfort from video terminals or usability studies of new devices)
- production lines (e.g., reducing musculoskeletal problems in assembly line workers).

Realistic digital humans (mannequins) in the virtual environment, for example *Jack*® (Bowman 2001) and *Ramsis*® (Geuss 2000; Vogt et al. 2005) to cite those most commonly used in both industrial and academic contexts, can help designers to perform human–machine interaction studies (see e.g., Colombo and Cugini 2005). These computer programs are widely used in the automotive field, where they were initially developed. When assessing the comfort of new car user packaging, these programs make it possible to analyze the joint angles that a subject can assume while driving, which can be related to the feeling of comfort (Porter and Gyi 1998). In such a virtual environment, the analysis of the interaction between a digital mock-up and the mannequin gives rise to a new form of experimentation.

These experiments in the virtual environment are usually time-consuming, so it is necessary to perform them in a rational and efficient way. Existing, well-known procedures for performing an ergonomic analysis (e.g., accessibility or usability analysis, comfort assessment) are based on the involvement of potential users selected on the basis of the mode, median and mean values and the limit percentiles for height. Using real or virtual experiments, it is possible to study the effects of anthropometric variation on ergonomic performance. Nevertheless, there is still a lack of a methodological framework for planning and analyzing virtual experiments aimed at improving product performance early on in the design process.

This chapter provides an overview of the work that the authors have carried out on what is termed the *robust ergonomic virtual design* (REVD) methodology. This approach has been applied to improve the comfort performances of driving seats and to identify innovative designs for seat adjustments in the virtual environment. This research work led to several articles (Barone et al. 2001, Barone and Lanzotti 2002; Barone et al. 2005; Barone and Lanzotti 2007; Lanzotti 2006; Lanzotti and Vanacore 2007a; Lanzotti and Vanacore 2007b). In Sect. 3.2, the methodological framework of REVD is presented, with particular emphasis placed on the definition of the performance indicator *weighted comfort loss* and the selection method for the experimental levels of the anthropometric noise factor *height*. An application is then described: a three-wheeled vehicle, where only the male population is considered as the target users. In Sect. 3.3, we focus on the most common situation, where the noise factor (the height of a potential user; female or male) is a mixture. For this case, we show how the correct experimental levels can be determined. A second application concerning mini-car user packaging is then presented. Section 3.4 presents the last phase of REVD, i.e., adjustment design, and the application of it to a new mini-car driving seat. Section 3.5 provides final comments and conclusions.

3.2 Aims and Phases of the Robust Ergonomic Virtual Design (REVD) Strategy

REVD is a statistics-based methodology for improving the ergonomic characteristics and finally the quality of a new product by reducing the sensitivity of its ergonomic performances to anthropometric variation.

Our starting point is the well-known robust design methodology (Taguchi 1987; Phadke 1989; Park 1996). Following this methodological path, and according to it, the REVD framework can be schematized as shown in Fig. 3.1.

The human–machine interface (HMI) design phase involves developing working prototypes of the system and fulfilling functional requirements by taking into account user interaction. The designer sketches several concepts, i.e., product architectures. Such concepts can, at this stage, be developed in a virtual environment as virtual prototypes that simulate some *key system ergonomic characteristics* (KSECs). This design phase is performed by exploiting engineering experience and knowledge, as well as simulation tools.

The parameter design phase involves identifying the main design parameters, and predicting and evaluating their optimal settings. In this context, attention is paid to improving system ergonomic robustness, i.e., insensitivity of the system to anthropometric variation. During the experimental phases, the designer can discover interaction effects that were not previously considered, and thus increase his/her scientific knowledge. Even if no "discovery" is made, the experimental phase at least contributes to improving the state of the art in his/her technological field.

The responses to analyze are the KSECs that were identified in the previous phase and that are measurable in the virtual environment. Design parameters are variables whose values can still be modified in this phase without any increase in manufacturing or usage costs. Designers select parameters that are able to significantly improve the KSECs (*key system ergonomic characteristics*). The effect of changing design parameter values is evaluated through experiments aimed at find-

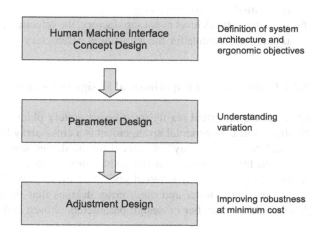

Fig. 3.1 The three main phases of robust ergonomic virtual design

ing the best design solution, otherwise termed the *design setting*, i.e., the best combination of design parameter values. In this phase, it is also necessary to identify the sources of variation—the so-called *noise factors*—and to understand their impact on the KSECs when the human–machine interaction is realistically evaluated. The systematic assessment of noise factors and their effects in product development is a growing research area (see, e.g., Johansson et al. 2006). Once identified, the noise factors should be opportunely introduced into the experimental arrangement (for a comprehensive account see, e.g., Wu and Hamada 2000). In REVD, the main source of noise is due to variation in the anthropometric dimensions of potential product users (e.g., vehicle drivers).

The adjustment design phase involves improving the system robustness still further if the optimal setting obtained in the parameter design phase does not guarantee a satisfactory performance level. The presence of an adjustment phase results from the inability to set any tolerance on the anthropometrical noise factor unless a decision is made to cut off some fraction of the potential user population.

Adjustment parameters can be identified in the parameter design phase when it is believed that it is technically viable to set them at minimum cost and that this does not negatively affect the safety and the cost of the product.

3.2.1 Key Steps of the Parameter and Adjustment Design Phases

Once the concept design for the human–machine interface has been completed, the key steps to follow according to the REVD methodology are:

1. Selection of the experimental design strategy to be performed in the virtual lab
2. Selection of the design parameters and levels
3. Identification of anthropometric noise factors and levels
4. Selection of the performance indicator (comfort loss)
5. Analysis of the experimental results and definition of the optimal settings for the main design parameters
6. If the robustness level is not satisfactory, the selection and definition of the functional requirements for the adjustment parameters

3.2.1.1 Selecting the Experimental Design to be Performed in the Virtual Lab

Experiments in virtual reality must be accurately planned. One effective and technically viable experimental arrangement is a cross-array (Phadke 1989; Park 1996), in which the "inner array" is defined by the design settings and the "outer array" is defined by the settings for the anthropometric noise factor values. Each design setting of the inner array is tested several times, as prescribed by the outer array.

The particular inner and outer array designs that should be chosen essentially depends on the number of design parameters chosen and the number of levels as-

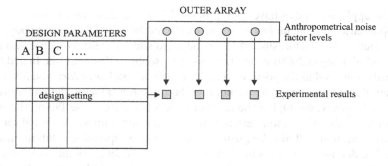

Fig. 3.2 Cross-array arrangement for experimentation in virtual reality

sociated with them. It also depends on the time and cost constraints imposed on experimenters (engineers/designers) during the development phase.

A pictorial representation of a cross-array is given in Fig. 3.2. The two plans for the *inner array* and *outer array* are crossed in the sense that each row of the inner array (design parameter level combination) is tested for each column of the outer array. The picture shows the case for a single anthropometric noise factor, which will be the case adopted here for reasons explained later.

3.2.1.2 Choice of Design Parameters and Levels

Design parameters depend on the product architecture and are defined during the concept design phase. In REVD, design parameters are essentially geometrical features (e.g., lengths, angles, etc.) of the design which are thought to affect ergonomic performance. It is interesting to investigate the extent to which they individually affect (main effects) the KSECs (*key system ergonomic characteristics*), but also to check whether there are any synergistic or antisynergistic interactions among these parameters. The choice of design parameter levels to be tested is mainly left to the judgment of designers/engineers, who will consider technical and economical constraints (in the sense that the adoption of a certain design parameter level should not significantly increase the manufacturing or usage cost).

3.2.1.3 Identification of Anthropometric Noise Factors and Levels

As already mentioned, the final objective of REVD is to find the product design setting that is the least sensitive (i.e., the most robust) to variations in the anthropometric characteristics of the subjects (men/women) who will interact with it. Therefore, in general, variations in the anthropometric characteristics of the target population are considered to be a noise factor in the robust design terminology.

In the applications that have been proposed by the authors up to now, only one anthropometric noise factor has been considered: the height of the human body. This is a synthetic noise factor, since it is a good surrogate for anthropometric variation (Reed and Flannagan 2000) within the particular scope of this analysis. In addition, it is easily managed in the posture prediction software packages *Jack* and *RAMSIS*.

When defining the outer array, more levels of the noise factor "height" must be chosen. Obviously, the higher the number of levels for the noise factor, the more representative the population variation will be, but the number of virtual lab experiments required will also be greater. Therefore, a compromise solution must be found. The definition of the noise factor levels in the outer array that is most commonly adopted is that proposed by Taguchi (see, e.g., Taguchi 1987), in which three levels are chosen: a central level corresponding to the expected value μ of the noise factor, and two levels at a distance $\pm\sigma\sqrt{3/2}$ from μ, which represent σ, the standard deviation of the noise factor. The responses obtained for these three noise factor levels are equally weighted. D'Errico and Zaino (1988) proposed an improvement of Taguchi's method that involved adopting three levels ($\mu - \sigma\sqrt{3}$, μ, $\mu + \sigma\sqrt{3}$) that were differently "weighted" ($1/6, 4/6, 1/6$, respectively). Their proposal was based on a better approximation of the continuous normal random variable (rv) when modeling the noise factor by a discrete three-mass-point rv.

The three levels proposed by D'Errico and Zaino (1988) are almost equivalent to the 5th, 50th and 95th height percentiles, which can be easily handled in the software packages *Jack* and *RAMSIS*.

3.2.1.4 Choice of the Performance Indicator: The Comfort Loss

In the ergonomic analysis of the HMI (human–machine interface), the use of a human mannequin in the virtual environment enables some defined KSECs to be measured. These KSECs do not fully explain the human feeling of comfort, since it is a complex and subjective structure, but they have the undoubted advantage of being objective and reproducible in the virtual reality experiments. For example, in the ergonomic analysis of new car user packaging, these programs provide the ability to evaluate the joint angles that a subject assumes while driving, which can be related to the human feeling of comfort (Porter and Gyi 1998, Barone and Curcio 2004). Joint angles are not the only way in which an objective measure of comfort can be derived. Another way is provided by the system of forces and pressures acting on the interface between the product and the user (Bubb and Estermann 2000).

In their investigation, based on a selected random sample representative of a population of western Europe drivers, Porter and Gyi (1998) measured the joint angles preferred by each subject when they were free to choose their preferred posture in a highly adjustable driving rig. Table 3.1 provides the summary statistics for a subset of preferred joint angles. These values concern only male drivers. However, the methodology is straightforward to extend to female drivers and eventually to a suitable mixture of the two, as will be shown later in this article.

Table 3.1 Summary statistics (values in degrees) of preferred joint angles
(source: Porter and Gyi 1998)

Joint angles	Min.	Preferred	Max.
Upper arm flexion	19	50	75
Elbow angle	86	128	164
Trunk–thigh angle	90	101	115
Knee angle	99	121	136
Foot–calf angle	80	93	113

We assume that the "preferred" joint angles in Table 3.1 are ideal targets for the designer. For simplicity of illustration, we start by considering only one joint angle (the knee angle). We define a "comfort loss" which is zero if the joint angle imposed by the design exactly matches the target value ($121°$ for the knee angle), while it increases as the imposed joint angle moves away from the target.

Due to the absence of quantitative knowledge about the discomfort experienced at the limits of the range, it is assumed that the comfort loss is the same value (conventionally placed equal to 1) at the extremes of each range given in Table 3.1. So the loss function we imagine for each joint angle is not symmetrical (Barone et al. 2001, Abdel-Malek et al. 2001). In order to test the most appropriate asymmetric loss function (for an example, see Spiring and Yeung 1998), Barone and Lanzotti (2002) considered several models of loss. Using this simplified measure of discomfort as well as human mannequins, we showed that analyses of discomfort data based on different models are substantially equivalent from the designer's point of view. Therefore, the simplest and most reasonable model, the quadratic asymmetric model, can be fruitfully applied. The quadratic asymmetric loss is an extension of the well known symmetric quadratic loss and is mostly used in tolerance design (Maghsoodloo and Li 2000). An analytical expression for it is:

$$L(y) = \begin{cases} \alpha(y - \tau)^2, & y \le \tau \\ \beta(y - \tau)^2, & y > \tau \end{cases} \tag{3.1}$$

where:

y \qquad is a generic joint angle value,

τ \qquad is the target value for the joint angle y,

α and β \qquad are constants. Imposing the loss value 1 for y_{min} and y_{max}, we get $\alpha = (y_{min} - \tau)^{-2}$ and $\beta = (y_{max} - \tau)^{-2}$.

In the applications presented in this article, we will always assume this model (3.1) for the loss function $L(y)$.

Now, it is evident that for a specific design setting, the value assumed by the joint angle depends on the anthropometric characteristics of the subject who will interact with that design (Pheasant 1996, Robertson and Minter 1996, Vogt et al. 2005). We are assuming that the height of the subject provides a good surrogate for the anthropometric variation. The height of the person interacting with the product is in fact a random variable (rv). We henceforth denote this rv "H."

Since the joint angle depends on H, it becomes a random variable, let's say $Y(H)$. Hence the loss function formulation becomes:

$$L(Y(H)) = \begin{cases} \alpha (Y(H) - \tau)^2, & Y(H) \leq \tau \\ \beta (Y(H) - \tau)^2, & Y(H) > \tau \end{cases} \tag{3.2}$$

We are interested in evaluating and maximizing the expected value of $L(Y(H))$, which can take the form:

$$E\{L(Y(H))\} = \int L(y(h)) \cdot f_H(h) \, dh \tag{3.3}$$

by considering H a continuous rv with a density $f_H(h)$.

If the continuous rv is replaced with a discrete three-mass-point rv, as specified in the previous subsection, then the expected loss can be rewritten as:

$$E\{L(Y(H))\} = \sum_{i=1}^{3} L(y(h_i)) \cdot w_i \tag{3.4}$$

where h_i are the three height values (5th, 50th, 95th percentiles) and w_i are the weights (probability masses respectively, $1/6$, $4/6$, $1/6$), set according to the D'Errico and Zaino approach.

By recalling Table 3.1, it is evident that a loss function for this case should consider all joint angles, and should thus have a multivariate structure.

Multivariate quadratic loss functions have been defined mainly for the symmetric case (see, e.g., Pignatiello 1993; Kuhnt and Erdbrügge 2004; and a recent review by Murphy et al. 2005). There, the loss function is defined as a quadratic form governed by a matrix of costs (known constants). The costs on the main diagonal of the matrix individually weight the distances of each of p responses from their targets. Furthermore, the off-diagonal elements of the matrix model the incremental losses incurred when pair of responses are simultaneously off-target. If the off-diagonal elements are placed equal to zero, the multivariate loss reduces to the sum of p univariate quadratic losses. Due to an absence of information, here we use a simplified model which assumes that all joint angles have the same importance to the comfort, and that off-diagonal cost elements are zero.

In this case, the loss function for each joint angle is:

$$L_j = L(Y_j(H)) = \begin{cases} \alpha_j (Y_j(H) - \tau_j)^2, & Y_j(H) \leq \tau_j \\ \beta_j (Y_j(H) - \tau_j)^2, & Y_j(H) > \tau_j \end{cases}, \quad j = 1, \dots, p \tag{3.5}$$

where $p = 5$, by referring to Table 3.1.

Under the assumption that off-diagonal cost elements are equal to zero, the total loss is given by:

$$L_{\text{tot}} = \sum_{j=1}^{p} L_j \tag{3.6}$$

and therefore the expected value of the loss is the sum of the expected values of the univariate losses, despite the correlation structure. So we define a total expected loss for the multivariate quadratic asymmetric case, and call it the weighted comfort loss (WCL):

$$\text{WCL} = \sum_{j=1}^{P} \sum_{i=1}^{3} L\left[y_j\left(h_i\right)\right] \cdot w_i . \tag{3.7}$$

3.2.1.5 Analysis of the Experimental Results

The WCL allows the anthropometric variability that affects joint angles to be accounted for in any fixed design setting; this makes the WCL the simplest approximate index for evaluating the overall comfort experienced by a potential user. Further analysis of the experimental results can be conducted on the basis of statistical methods. It can be a more or less standard analysis, depending on the experimental arrangement adopted and the degree of mathematical precision. Main effects analysis and Pareto diagrams are useful and simple tools for evaluating the design parameters during explorative experimental phases.

3.2.2 A First Application: Three-Wheeled Vehicle

In this first application, we show the results obtained during the development phase for an electrically powered three-wheeled vehicle permitting city center mobility. Figure 3.3a shows a digital mock-up of the vehicle concept. It was developed using a computer-aided styling program and then transferred to the *Jack®* software. Figure 3.3b shows how the mannequin can be accommodated on the vehicle mock-up by imposing several constraints: the hands grasp the handlebar, the feet are located at a fixed point on the foot-board, the rear is located at a fixed point on the seat, and the mannequin's back is held vertical.

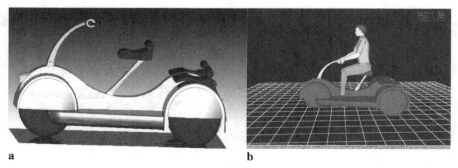

a   b

Fig. 3.3 a Digital mock-up of the vehicle concept proposed by the designer Francesco Fittipaldi. **b** The mannequin *Jack®* accommodated in the mock-up in the virtual environment

Table 3.2 Description of design parameters and levels

Parameter		Levels 0	1	2
A	Handlebar arc length (mm)	500	600	700
B	Seat arm length (mm)	430	445	460
C	Handlebar angular position (deg.)	−15	0	15
D	Seat arm angular position (deg.)	−15	0	15
E	Seat angular inclination (deg.)	−10	0	10

In this early development phase, an aim of the designer is to anticipate the ergonomic characteristics of the vehicle as accurately as possible by finding the geometrical features that have the most effect on the driver's comfort and their optimal settings, which should then be frozen during the following development phases. This activity can also result in the proposal of adjustment factors, i.e., design features that can be regulated by the end-user, taking into consideration safety and cost constraints.

The six key methodological steps described in Sect. 3.2 can be followed here. The designer must choose the design parameters and their levels. Five parameters were identified and judged as potentially affecting driver comfort. They are described in Table 3.2 and depicted in the schematic drawing of Fig. 3.4.

For each design parameter, three levels were chosen; level 1 represents the initial setting, i.e., a setting tentatively chosen by the designer based on his/her experience only.

During the virtual experimentation, for each design setting it is necessary to:

* Build the digital mock-up of the vehicle corresponding to the design setting
* Import the obtained digital mock-up into the virtual environment
* Choose the anthropometric dimensions of the mannequin
* Accommodate the mannequin in the vehicle
* Analyze the joint angles

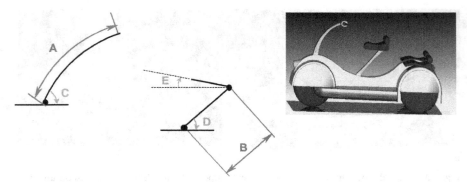

Fig. 3.4 Scheme of the adopted design parameters

These operations cannot be performed automatically. Each run is time-consuming and requires the work of at least one person. Furthermore, the experimenters need to account for all conditions to ensure reproducibility.

In order to find the optimal design setting, a cross-array was planned and used. A three-level 3^{5-2} fractional factorial design was adopted as the inner array (Wu and Hamada 2000), while the anthropometric noise factor "height" was set at the 5th, 50th and 95th percentiles, since the product was initially intended only for men. The experimental plan and summary results in terms of WCL (weighted comfort loss) are shown in Table 3.3. To get a benchmark, the WCL was calculated at the initial setting (all factors coded at level 1), which is not included in the inner array. For this, WCL0 = 2.83.

From the data in Table 3.3, it is possible to estimate the main effects of the design parameters on the WCL. They are also presented graphically in Fig. 3.5. Recall that each point in the graph represents an average of nine different experimental conditions, this being a strength of this experimental design.

Table 3.3 Experimental design of and results from the first parameter design phase for the three-wheeled vehicle (DS = design setting)

DS	A	B	C	D	E	WCL	DS	A	B	C	D	E	WCL	DS	A	B	C	D	E	WCL
1	0	0	0	0	0	23.28	10	1	0	0	1	1	5.38	19	2	0	0	2	2	3.04
2	0	0	1	1	2	2.15	11	1	0	1	2	0	4.20	20	2	0	1	0	1	19.64
3	0	0	2	2	1	2.76	12	1	0	2	0	2	10.75	21	2	0	2	1	0	13.45
4	0	1	0	1	2	2.65	13	1	1	0	2	0	2.85	22	2	1	0	0	1	17.27
5	0	1	1	2	1	2.21	14	1	1	1	0	2	8.40	23	2	1	1	1	0	8.43
6	0	1	2	0	0	18.85	15	1	1	2	1	1	4.09	24	2	1	2	2	2	2.23
7	0	2	0	2	1	3.62	16	1	2	0	0	2	4.04	25	2	2	0	1	0	5.56
8	0	2	1	0	0	15.57	17	1	2	1	1	1	1.78	26	2	2	1	2	2	4.09
9	0	2	2	1	2	2.32	18	1	2	2	2	0	3.32	27	2	2	2	0	1	9.35

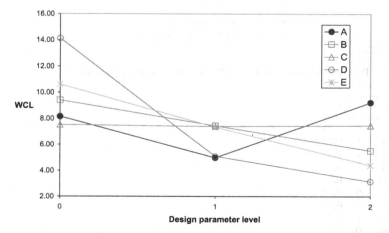

Fig. 3.5 Design parameter main effects on the weighted comfort loss

From Fig. 3.5, it is clear that the handlebar arc length (factor A) has an evident quadratic effect, while the most evident linear effect is given by the seat arm angular position (factor D), for which the best level tested is level 2. Seat arm length and seat angular inclination (factors B and E) have moderate linear effects. Handlebar angular position (factor C) results are indifferent (at least across the tested range). The best experimental design setting is the 17th, for which we have calculated a WCL = 1.78—already a 37% improvement over the initial setting. However, the best experimental design setting is different from the expected optimal setting (factor A at level 1, B at level 2, C at level 0, D at level 2, E at level 2), as can be deduced by selecting the level associated with the lowest WCL value for each factor from Fig. 3.5.

To confirm the previous results, a second experimental phase was performed (see Tables 3.4 and 3.5). Factor A was fixed at the previous level, 1, factor C, the setting of which appeared to be largely irrelevant, was fixed at level 2, the most convenient one. Factor D was fixed at the previous level, 2, and was proposed as an adjustment factor. The second inner array tested was a central composite design for the two factors B and E, enabling the response surface to be estimated. The levels for factor B were chosen to investigate whether WCL decreases for values that are higher than the best one found in the previous phase. The levels for factor E were chosen to

Table 3.4 Description of the design parameters and levels for the second experimental phase

						Dimensions
Fixed parameters						
A			600			mm
C				15		degrees
D				15		degrees
Design parameters	Levels					
	$-\alpha$	-1	0	1	α	
B	440	446	460	474	480	mm
E	7.5	8.2	10	11.8	12.5	degrees

Table 3.5 Experimental design of and results from the second experimental phase (DS = design setting)

DS	B	E	WCL
1	-1	-1	0.74
2	1	-1	0.49
3	-1	1	0.68
4	1	1	0.87
5	$-\alpha$	0	0.50
6	a	0	0.60
7	0	$-\alpha$	0.69
8	0	a	0.73
9	0	0	0.56

check whether the WCL minimum occurred within the range tested previously, due to engineering constraints that required the maximum level of this factor to increase.

The predicted minimum WCL lies at the point with coordinates $(-0.8109, 0.7454)$, corresponding to the levels B $= 361$ mm and E $= 8.79$ degrees. The predicted value of WCL is 0.46. The point was tested and the prediction error was negligible. The final improvement attained in WCL was 84% ($|0.46 - 2.83|/2.83$). This application showed that, although the initial setting of design parameters based on designer experience was good, the application of this methodology made it possible to attain significant improvements with limited experimental effort (two experimental sessions performed in about two working days).

3.3 Anthropometric Noise Factor for a Mixture Population

The approach proposed by D'Errico and Zaino to model the anthropometric noise factor in the cross-array experiment is unsuitable when the designer wishes to consider a mixture of men and women for the potential population of product users. In this case, the anthropometric noise factor is a mixture of two rvs. We consider the rv "height" H to be a mixture of the two normal rvs H_w and H_m. We then pose: $\mu_w = E\{H_w\}$, $\mu_m = E\{H_m\}$, $\sigma_w^2 = \text{Var}\{H_w\}$, $\sigma_m^2 = \text{Var}\{H_m\}$.

The probability density function (pdf) of H is simply obtained by weighting the pdfs of H_w and H_m:

$$f_H(h) = k \cdot f_w(h) + (1-k) \cdot f_m(h) \tag{3.8}$$

where k is the mix coefficient (proportion of women in the target population), and the parameters are $\mu_w = 1.627$; $\sigma_w = 0.069$; $\mu_m = 1.755$; $\sigma_m = 0.068$.

The choice of levels for the anthropometric noise factor "height" in the REVD experiment poses a problem when approximating the mixture H by a discrete rv assuming only a very limited number of prefixed values. We henceforth consider the case in which four levels are chosen for the noise factor; such values are denoted by h_i ($i = 1, \ldots, 4$). In particular, we choose the values h_1 (corresponding to the 5th percentile of H_w), h_2 (to the 50th percentile of H_w), h_3 (to the 50th percentile of H_m), and h_4 (to the 95th percentile of H_m). This choice is reasonable and can be adopted mainly for practical purposes (these percentiles are easily set in the packages *Jack* and *RAMSIS*). However, Barone and Lanzotti (2007) also showed the methodological reasons for the optimal nature of this choice (four noise factor levels).

To obtain equivalence for the two probability distributions (the continuous mixture and the discrete approximation), we equate the first three absolute moments of the two distributions. Furthermore, we impose the necessary constraint on the sum of probabilities for the discrete rv (Barone and Lanzotti 2007). For the parameters defined, we have $h_1 = 1.513$; $h_2 = \mu_w = 1.627$; $h_3 = \mu_m = 1.755$; $h_4 = 1.866$. By solving the system, we get: $w_1 = 0.049$; $w_2 = 0.316$; $w_3 = 0.511$; $w_4 = 0.124$. It should be noted that the discrete approximation is parametrically defined by the mix coefficient k.

For this case of a mixed population, formula (3.7) is generalized to the following:

$$\text{WCL} = \text{WCL}(k) = \sum_{j=1}^{p} \sum_{i=1}^{4} L[y_j(h_i)] \cdot w_i(k) . \tag{3.9}$$

3.3.1 A Second Application: Mini-Car User Packaging

As a second application, we illustrate the results of work carried out in the development of mini-car user packaging. Figure 3.6a shows the vehicle concept, while Fig. 3.6b indicates the seven design parameters chosen for the experiments in the virtual lab. These parameters and their levels were selected based on designer experience and SAE guidelines (SAE = Society of Automotive Engineers).

Table 3.6 provides details on the chosen design parameters and the levels that were tested.

Once the design parameters and their levels had been defined, the experiments were performed according to a cross-array arrangement. A fractional 2_{IV}^{7-3} factorial (Wu and Hamada 2000) consisting of sixteen design settings was chosen as the inner

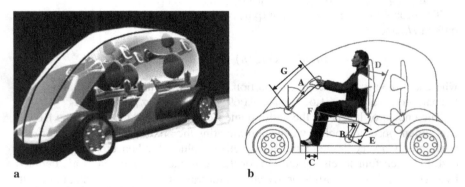

a   b

Fig. 3.6a,b REVD of mini-car user packaging (the mini-car model was designed by F. Fittipaldi)

Table 3.6 Description of design parameters and levels for the experimental phase (coded as "-1" and "$+1$")

Parameter		Experimental levels	
A	Steering wheel angular position (deg)	40	50
B	Seat arm angular position (deg)	20	40
C	Foot position on footboard (mm)	5	15
D	Seat back inclination (deg)	0	5
E	Seat arm length (mm)	30	35
F	Seat cushion inclination (deg)	0	5
G	Steering wheel arm length (mm)	55	65

Table 3.7 Experimental design of and results for the mini-car user packaging (DS = design setting)

DS	A	B	C	D	E	F	G	WCL (k = 0.25)	WCL (k = 0.50)	WCL (k = 0.75)
1	−1	−1	−1	−1	−1	−1	−1	5.37	4.83	4.29
2	1	−1	−1	−1	1	−1	1	6.55	6.21	5.87
3	−1	1	−1	−1	1	1	−1	1.67	1.38	1.10
4	1	1	−1	−1	−1	1	1	6.95	6.39	5.83
5	−1	−1	1	−1	1	1	1	8.09	7.57	7.05
6	1	−1	1	−1	−1	1	−1	9.42	8.88	8.33
7	−1	1	1	−1	−1	−1	1	7.72	6.98	6.23
8	1	1	1	−1	1	−1	−1	3.50	2.97	2.44
9	−1	−1	−1	1	−1	1	1	5.61	5.19	4.77
10	1	−1	−1	1	1	1	−1	3.82	3.67	3.52
11	−1	1	−1	1	1	−1	1	1.87	1.74	1.61
12	1	1	−1	1	−1	−1	−1	2.30	2.06	1.81
13	−1	−1	1	1	1	−1	−1	6.56	6.14	5.73
14	1	−1	1	1	−1	−1	1	12.07	11.38	10.69
15	−1	1	1	1	−1	1	−1	4.34	3.80	3.26
16	1	1	1	1	1	1	1	4.55	4.04	3.52

array. This is presented in Table 3.7. The last three columns of the table show the WCL values obtained using formula (3.9).

Figure 3.7 shows the estimated main effect plot for the mix coefficient $k = 0.5$. The graph shows that the most important effects are: B (seat arm angular position), C (foot position on footboard), E (seat arm length), and G (steering wheel arm length). Factor F (seat cushion inclination) has absolutely no influence.

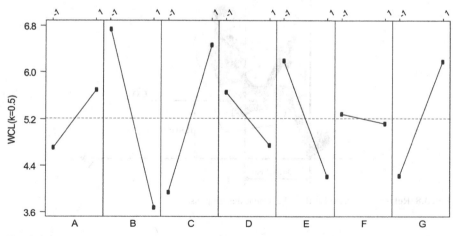

Fig. 3.7 Estimated effects of the design parameters on WCL (mix coefficient $k = 0.5$)

3.4 Adjustment Design of a New Mini-Car Driving Seat

An application of the last step of the REVD methodology is now presented, concerning the concept design of a new mini-car driving seat. After the definition of the mini-car user packaging (*see* Lanzotti 2008), involving steps 1 to 5 (Sect. 3.2.1), the adjustment design phase is performed to improve system comfort. This phase starts with the optimal settings obtained at the end of the parameter design phase and aims to further improve the robustness of the system to anthropometric variability.

First, to set the functional requirements for the adjustment parameters of the driving seat, it is necessary to establish a reference system. The origin of the parameter space is the *H-point*. Its position is defined by the coordinates H30, the vertical distance between the seating reference point (SgRP) and the heel point, and L53, the horizontal distance between SgRP and the heel point (see Fig. 3.8). Their initial values, $(58.2, 50)$ (expressed in cm), were obtained as a result of the parameter design phase.

The experimental phase aims to explore the new parameter space with classical techniques ("steepest ascent"). This space has its origin at the coordinate (L53, H30), equal to $(58.2, 50)$. The exploration is made in three steps. The first two steps are carried out using two full-factor factorial experimental design with central point experiments. The third step is carried out using a central composite design, with the central point occurring at coordinates $(8, -2)$. The results are reported in Table 3.8.

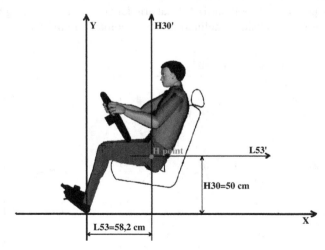

Fig. 3.8 Reference system for the adjustment design phase

Table 3.8 Comfort loss evaluated with a central composite design (CCD)

Run	A (H30)	B* (L53)	Female		Male	
			5th	50th	50th	95th
0	0.00	10.00	11.61	6.33	2.47	1.72
1	−4.00	10.00	8.24	6.20	2.76	3.29
2	−2.00	5.17	4.15	3.50	3.99	4.34
3	−2.00	10.83	10.34	6.48	2.90	2.18
4	0.83	8.00	7.83	4.76	2.41	2.40
5	−4.00	6.00	6.36	4.04	2.96	3.24
6	−2.00	8.00	5.76	3.32	2.30	2.78
7	0.00	6.00	6.31	4.03	3.15	3.36
8	−4.83	8.00	5.76	4.36	3.12	3.63

3.4.1 Sensitivity of Comfort Loss to Mix Variation

Figure 3.9 shows isocomfort curves, and uses different colors to highlight the decreasing comfort around the new optimal area (the blue one, where the comfort loss is less than 4).

The optimal area is defined by a comfort loss that is 10% higher than the minimum value and by an absolute value of less than 4, empirically fixed on the basis of the conventional limit value for the loss (equal to 1) for all joint angles (in number of 12). The isocomfort curves show that the optimal posture of the mannequin, defined at the end of the parameter design phase, can be significantly improved (about 50%).

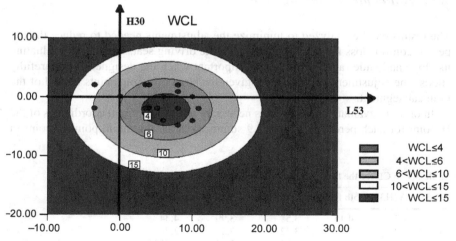

Fig. 3.9 Isocomfort curves based on WCL(50) for the 50% population mix

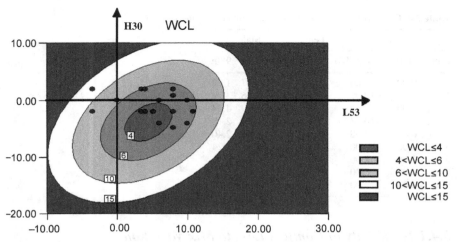

Fig. 3.10 Isocomfort curves based on WCL(75) for a 75% population mix

Figure 3.10 shows how isocomfort curves vary with the mix coefficient. This means that the system is not robust in the widest possible sense, since the optimal setting depends on the population mix. This variability could be reduced if the product is designed only for a specific target (e.g., for female users or a prefixed mix); when this target is not specified, the robustness can be improved only by introducing adjustment parameters.

3.4.2 Adjustment Optimization

The results can be analyzed to minimize the adjustments needed to reduce the expected comfort loss for the population. For the driving seat, this means evaluating its slope and slide range. This step is important when the designer must carefully choose the adjustments to introduce, given the strict cost constraints typical of the mini car segment.

In order to evaluate the slope, it is necessary to fit the optimal coordinates of the H-point for each percentile. Table 3.9 summarizes the minimum point of comfort

Table 3.9 WCL for the four percentiles used in the virtual experiments

X(L53)	Y(H30)	5th female	50th female	50th male	95th male
5.2	−2	4.15	3.5	3.99	4.34
8	−2	5.76	3.32	2.3	2.78
10	0	11.61	6.33	2.47	1.72

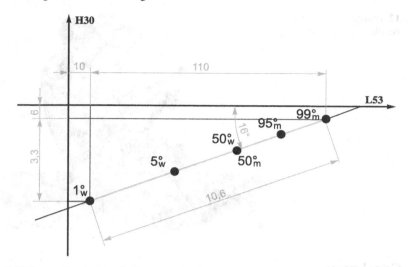

Fig. 3.11 Innovative design of adjustment requirements

loss for each percentile. By applying the weighted least square technique, it is possible to estimate the regression line for each population mix. As an example, the estimation procedure can be repeated for the quantiles and the median (25%, 50% and 75%), in order to compare the slopes.

The slope ranges from 6° (for the 25%) to 28° (for the 75%). Thus, a device that is able to satisfy this range can maximize the comfort. Otherwise, once the population mix has been fixed, we can design an innovative device with a slope that minimizes the comfort loss. At this step, the second adjustment parameter to design is the slide range. In order to minimize the comfort loss of potential users, the 1st female and 99th male percentiles need to be introduced into the experiments performed at this step (Lanzotti 2008b).

Figure 3.11 shows the results for the optimal setting of the slide range. The value of 110 mm enables the driver cabin to be the smallest, a useful advantage in the mini-car segment.

These results are good if compared with those obtained by UMTRI (University of Michigan Transportation Research Institute) and implemented in Jack. The horizontal slide range proposed by UMTRI is 146 mm—more than that proposed here (Reed et al. 1999, 2000).

Figure 3.12 shows the positions of the six percentiles of the human mannequins. It is useful to check the driver cabin and the accessibility to the dashboard.

Fig. 3.12 Positioning of the
six percentiles

3.5 Conclusions

This chapter presented the methodological framework of REVD (robust ergonomic virtual design), which involves HMI (human–machine interface) concept design and robustness improvement. The HMI is tested using a cross-array experimental design in a virtual environment. The chapter has focused on the pre-design phase of the outer array, when the experimenter has to select the levels of the anthropometric noise factor. In this case, the noise factor is a mixture of two Normal variables, corresponding to the male and female components of a target population of potential product users. For this problem, an efficient solution was applied that saves 33% of experimental effort compared with a straightforward generalization of D'Errico and Zaino's approach to the discretization of noise factors.

Finally, for cases where the system response—in terms of discomfort minimization for potential users—is not yet *robust* enough to the anthropometric noise factor, the authors have briefly illustrated, as a guideline, how the most significant design factors can be identified and then adjusted in accordance with the anthropometric particulars of the potential user. In fact, no restriction (i.e., tolerance) can be placed on the anthropometric noise factors unless a decision is made to design out a fraction of the population of potential users.

As a beautiful synthesis of artistic and statistical designer cooperation, Fig. 3.13 shows sketches made by the Italian designer Riccardo Dalisi, who was involved in this project. Starting from the optimal definition of the main dimensions of an office chair, represented by the continuous line in Fig. 3.13 and obtained using the REVD approach by postgraduate and Ph.D. students working for their theses, Riccardo Dalisi has creatively used these lines while giving his concept design the right proportions.

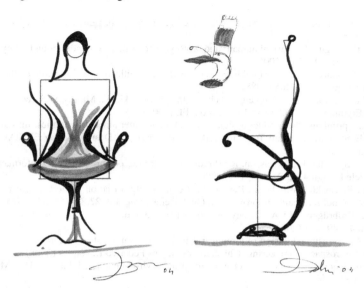

Fig. 3.13 REVD of an office chair. The artistic side of the research work (designed by Riccardo Dalisi, private collection)

Acknowledgements This work was financially supported by the project (PRIN) "Statistical design of continuous product innovation," funded by the Italian Ministry of University and Research. The authors, who mutually acknowledge an equal contribution to this work, wish to thank Francesco Carbone and all Ph.D. and postgraduate students who contributed to the experimental work, and the designers Francesco Fittipaldi and Riccardo Dalisi for their creativity and enthusiastic involvement in this research.

References

Barone, S., Curcio, A.: A computer-aided design-based system for posture analyses of motorcycles. J. Eng. Des. **15**(6), 581–595 (2004)

Barone, S., Carbone, F., Lanzotti, A.: Progettazione del posto guida di una minicar basata su esperienza del designer e sperimentazione virtuale. In: Proc. Congreso Internacional Conjunto XVII INGEGRAF—XV ADM, Sevilla, Spain, ISBN 84-96377-41-5, 1–3 June (2005)

Barone, S., Fittipaldi, F., Lanzotti, A.: Improving comfort of a new city vehicle by means of parameter design in virtual environment. In: Proc. 1st Annual Conf. European Network for Business and Industrial Statistics, Oslo, Norway, 17–18 Sept. (2001)

Barone, S., Lanzotti, A.: Quality engineering approach to improve comfort of a new vehicle in virtual environment. Proceedings of the ASA, statistical computing section. American Statistical Association, Alexandria, VA (2002)

Barone, S., Lanzotti, A.: On the treatment of anthropometrical noise factors in robust ergonomic design. In: Dall'Idea al Prodotto: La Rappresentazione Come Base per lo Sviluppo e l'Innovazione. Edizioni ETS, Pisa, pp 359–367, ISBN 978-884671841-9 (2007)

Bowman, D.: Using digital human modeling in a virtual heavy vehicle development environment. In: Chaffin, D. (ed.): Proceedings of the Digital Human Modeling Conference. SAE, Munich, pp 343–353 (2001)

Bubb, H., Estermann, S.: Influence of forces on comfort feeling in vehicles. SAE Paper 2000-01-2171 (2000)

Colombo, G., Cugini, U.: Virtual humans and prototypes to evaluate ergonomics and safety. J. Eng. Design 16(2), 195–207 (2005)

D'Errico, J., Zaino, R.A.: Statistical tolerancing using a modification of Taguchi's method. Technometrics 30(4), 397–405 (1988)

Dainoff, M.J.: Ergonomics of seating and chairs. In: Karwowski, W., Marras, W.S. (eds.): Occupational Ergonomics. CRC Press, Boca Raton, FL (2003)

Geuss, H.: Optimizing the product design process by computer aided ergonomics. In: Landau, K. (ed.): Ergonomic Software Tools in Product and Workplace Design. Verlag ERGON GmbH, Stuttgart (2000)

Hanson, L., Sperling, L., Akselsson, R.: Preferred car driving posture using 3-D information. Int. J. Vehicle Design 42(1–2), 154–169 (2006)

Johansson, P., Chakhunashvili, A., Barone, S., Bergman, B.: Variation mode and effect analysis: a practical tool for quality improvement. Qual. Reliab. Eng. Int. 22(8), 865–876 (2006)

Kuhnt, S., Erdbrügge, M.: A strategy of robust parameter design for multiple responses. Stat. Model. 4, 249–264 (2004)

Lanzotti, A.: Adjustment design of a minicar driving seat. (unpublished research report). Department of Aerospace Engineering, University of Naples Federico II, Naples (2008)

Lanzotti, A.: Robust design of car packaging in virtual environment. Int. J. Interact. Des. Manufact. 2(1), 39–46 (2008)

Lanzotti, A., Vanacore, A.: An efficient and easy discretizing method for the treatment of noise factors in robust design. Asian J. Qual. 8(3), ISSN 1598-2688 (2007)

Levy, M.S., Wen, D.: BLINEX: A bounded asymmetric loss function with application to Bayesian estimation. Commun Stat Theory Methods 30, 147–153 (2001)

Maghsoodloo, S., Li, M.C.: Optimal asymmetric tolerance design. IIE Trans. 32, 1127–1137 (2000)

Murphy, T.E., Tsui, K.-L., Allen, J.K.: A review of robust design methods for multiple responses. Res. Eng. Des. 16, 118–132 (2005)

Park, S.H.: Robust Design and Analysis for Quality Engineering. Chapman & Hall, London (1996)

Phadke, S.M.: Quality Engineering Using Robust Design. Prentice-Hall, Upper Saddle River, NJ (1989)

Pheasant, S.: Bodyspace: Anthropometry, Ergonomics and the Design of the Work. Taylor & Francis, London (1996)

Pignatiello, J.: Strategies for robust multiresponse quality engineering. IIE Trans. 25(3), 5–15 (1993)

Porter, M., Gyi, D.E.: Exploring the optimum posture for driver comfort. Int. J. Vehicle Des. 19(3), 255–266 (1998)

Reed, M.P., Flannagan, C.A.C.: Anthropometric and postural variability: limitations of the boundary manikin approach. SAE Paper 2000-01-2172 (2000)

Reed, M.P., Manary, M.A., Flannagan, A.C., Schneider, W.L.: New concept in vehicle interior design using aspect. SAE Tech. Paper 1999-01-0967 (1999)

Reed, M.P., Manary, M.A., Flannagan A.C., Schneider, W.L.: Comparison of methods for predicting automotive driver posture. SAE Tech. Paper 2000-01-2180 (2000)

Robertson, S.A., Minter, A.: A study of some anthropometric characteristics of motorcycle riders. Appl. Ergonom. 27(4), 223–229 (1996)

Seo, H.S., Kwak, B.M.: Efficient statistical tolerance analysis for general distributions using three-point information. Int. J. Prod. Res. 40(4), 931–944 (2002)

Spiring, F.A., Yeung, A.S.: A general class of loss functions with industrial applications. J. Qual. Technol. 30(2), 152–162 (1998)

Vogt, C., Mergl, C., Bubb, H.: Interior layout design of passenger vehicles with RAMSIS. Hum. Factors Ergonom. Manufact. 15(2), 197–212 (2005)

Taguchi, G.: System of Experimental Design. Kraus, New York (1987)

Wu, C.F.J., Hamada, M.: Experiments. Wiley, New York (2000)

Chapter 4
Computer Simulations for the Optimization of Technological Processes

Alessandro Baldi Antognini, Alessandra Giovagnoli,
Daniele Romano, and Maroussa Zagoraiou

Abstract This chapter is about experiments for quality improvement and the innovation of products and processes performed by computer simulation. It describes familiar methods for creating surrogate models of simulators (emulators), with particular reference to Kriging interpolation, and some new ways of fitting the models to the simulated data.

It also deals with the advantages of computer experiments performed sequentially, and with computer experiments in which some of the random noise factors that affect the output of a process are simulated stochastically.

As an example, an application to the integrated parameter and tolerance design of a high-precision space-measuring instrument is illustrated, and other potential applications are also mentioned.

Alessandro Baldi Antognini
Department of Statistical Sciences, University of Bologna
e-mail: a.baldi@unibo.it

Alessandra Giovagnoli
Department of Statistical Sciences, University of Bologna
e-mail: alessandra.giovagnoli@unibo.it

Daniele Romano
Department of Mechanical Engineering, University of Cagliari
Piazza d'Armi, 1, 09123 Cagliari, Italy
e-mail: romano@dimeca.unica.it

Maroussa Zagoraiou
Department of Statistical Sciences, University of Bologna
e-mail: maroussa.zagoraiou@unibo.it

4.1 Introduction

4.1.1 Importance of Computer Simulation

The importance of experimenting for quality improvement and innovation of products and processes is now very well known: "experimenting" means to implement significant and intentional changes with the aim of obtaining useful information. In particular, the majority of industrial experiments have two goals:

- To quantify the dependence of one or more observable response variables on a group of input factors in the design or the manufacturing of a product, in order to forecast the behavior of the system in a reliable way.
- To identify the level settings for the inputs (design parameters) that are capable of optimizing the response.

The set of rules that govern experiments for technological improvement in a physical set-up are now comprehensively labeled "DoE." In recent years, the use of experimentation in engineering design has received renewed momentum through the utilization of **computer** experiments (see Sacks et al. 1989, Santner et al. 2003), which has been steadily growing in the last two decades. These experiments are run on a computer code implementing a simulation model of a physical system of interest. This enables us to explore the complex relationships between input and output variables. The main advantage of this is that the system becomes more "observable," since computer runs are generally easier and cheaper than measurements taken in a physical set-up, and the exploration can be carried out more thoroughly. This is particularly attractive in industrial design applications where the goal is system optimization.

4.1.2 Simulators and Emulators

The request that the simulator should be accurate in describing the physical system means that the simulator itself may be rather complex. In general a "simulator" will consist of a set of many linear or nonlinear, ordinary and/or differential simultaneous equations, whose solutions may not be amenable to analytical expression. Furthermore, the number of input factors may be large, so runs may be expensive and/or time-consuming. This has led to the use of surrogate models (emulators), i.e., simpler models which represent a valid approximation of the original simulator. These emulators are statistical interpolators built from the simulated input–output data. Predictions at untried points, most useful in the case of expensive simulations, are made by the surrogate models.

Familiar methods for creating surrogate models are the response surface methodology and the Kriging interpolator. The latter is a technique where the deterministic response of the computer is regarded as a realization of a random process whose correlation function can be shaped in such a way as to reflect the characteristics of the response surface (for more details, see Santner et al. 2003). This and other types of emulators are reviewed in Sect. 4.2.

The existing methods of fitting a Kriging model generally start with a space-filling design, e.g., a latin hypercube sampling scheme. A novel method based on the theory of optimal designs has been introduced by Zagoraiou and Baldi Antognini (2008), and will be explained in Sect. 4.2.

4.1.3 Sequential Computer Experiments

Originally introduced by Box and Wilson (1951) in the context of statistical quality control and optimization of engineering processes/products, sequential experimentation includes a vast class of statistical procedures which employ previously collected data in order to modify the trial as it goes along. Because of their greater flexibility and efficiency than classical fixed-sample procedures (see for instance Siegmund 1985 and Ghosh and Sen 1991), sequential procedures are now widely used in the great majority of applied contexts, such as biomedical practice, physical research, and industrial experimentation. In this field in particular, where economic demands (time and cost savings) may be crucial, the ability to exploit the newly acquired data during the trial may help to detect potential defects, inefficient settings, dangerous or ineffective treatments, and enable designers to pursue continuous quality improvement.

There is a fair amount of recent literature on this topic (Romano 2006): existing sequential methods generally start with a space-filling design (a latin hypercube sampling design or a *maximin* design); then, once the sequential procedure is activated, the estimates for the parameters of the unknown quantities process are recursively updated at each step or, to reduce the computational burden, after a few steps. Adaptive designs for optimization in the response surface methodology framework are available that iteratively reduce the design space. Wang et al. (2001) discard regions with large values of the objective function at each modeling–optimization iteration, while Farhang-Mehr and Azarm (2003) and Lin et al. (2004) use maximum entropy as the criterion. Other criteria for successively reducing the design space include move limits (Wujek and Renaud 1998) and trust regions (Alexandrov et al. 1998). Van Beers and Kleijnen (2005) use bootstrap to estimate the prediction variance at untried sites and then choose the point with the maximum prediction variance as the next input.

In Sect. 4.3, the principles of sequential experimentation are illustrated, followed by some recent proposals for sequential procedures. A new virtual experiment for a Kriging model by Baldi Antognini and Pedone is introduced which is randomized, thus making it possible to overcome a typical problem, namely that algorithms tend to linger for a long time around local maxima/minima. The new method extends the one by Kleijnen and Van Beers (2004) and has turned out to be very efficient for small samples too, reaching the same degree of precision with 40% fewer observations.

4.1.4 Stochastic Simulators

A major problem with most simulators is that they use a deterministic code. This choice is too restrictive. In many instances, some important input variables are random in the real process. Typical examples are the design of robust engineering systems, where parameters that cannot be controlled by the designer (like external temperature and manufacturing errors) act randomly, and the management and control of queuing systems (for instance production and telecommunication facilities), where arrival and service times (of parts, phone calls) are random. In such cases, the simulation code, e.g., a finite element code for a new product (Romano et al. 2004) or a discrete event simulator for inventory systems (Bashyam and Fu 1998) may be random too: this appears to be the natural tool for transmitting the distribution of noise input to the output. Besides, the construction of a computer simulator of a complex system often has some degree of uncertainty. This involves several decisions about different modeling options, numerical algorithms, and the assignment of values to physical and numerical parameters. Since there are often no clear-cut best decisions, a huge number of deterministic computer codes are compatible with the same physical system. The same is true of course of stochastic simulators, but to a lesser degree. Thus (Romano and Vicario 2002a, 2002b) random simulation again seems preferable, and it is now largely employed in several technological and scientific areas. One practical consequence is that the rationale for using standard statistical tools is restored. Thus regression analysis, for instance, can be safely used for prediction.

In Sect. 4.4, a modified protocol for conducting robust design studies on the computer is described, which extends the dual response surface approach (Giovagnoli and Romano 2008). The suggested protocol utilizes stochastic simulation. It is characterized by a different treatment of the noise factors, some of which are considered random, as they are in the real process. The proposed model includes both the *crossed array* and the *combined array* as special cases. As is well known, crossed arrays are the experimental plans suggested by Taguchi and Wu (1980): some level combinations of the control factors (inner array) are chosen and tested across some level combinations of the random factors (outer array), which in the experimental setting are selected by the experimenters and are under their control; this type of experiment permits the fitting of two separate response surfaces: one for the response and one for the log-variance. An alternative is a combined array experiment (Vining and Myers 1990; Welch et al. 1990), which fits a second-degree polynomial regression model—containing both control factors and their interactions with the noise variables—with fixed levels of both the controls and the noises.

When the design scope is extended to the specification of allowable deviations of parameters from nominal settings (tolerances), an integrated parameter and tolerance design problem arises. Here the additional objective of minimizing the production costs needed to fulfil tolerance specs will compete with the minimum variance objective. This method can be beneficial to the solution of an integrated parameter and tolerance design problem, since it adds standard deviations of **internal** noise factors as controls in the experiment and simulates the internal noise accordingly.

The method appears to be particularly well suited to complex measurement system design. The main quality requirements of a measure, namely lack of bias and precision, are strictly linked to the mean and variance of the measurement. In Giovagnoli and Romano (2008), we describe how the method was applied to the design of a high-precision measuring instrument, an optical profilometer. In Sect. 4.4, this case study is recalled briefly to show that the same method can be applied in a wide range of other possible contexts.

4.2 Construction of the Emulators

A common approach used to deal with the problem of complex input–output relationships exhibited by the simulation model is to construct an emulator, also called a surrogate. Since emulators are models of the original simulator, which is itself a model of reality, they are often called metamodels. The goal of using a surrogate model is to provide a smooth relationship of potentially high fidelity to the true function with the advantage of quick computational speed instead of the time-consuming runs associated with the computer code. This section provides guidelines for the construction of an emulator and a brief overview of many different emulators; it does not aim to explain each type in detail but rather to underline the wide variety of approximating models available in the literature. The Kriging methodology is discussed in a thorough way due to its importance as a tool for accurate predictions.

4.2.1 A Protocol for Creating Emulators

Kleijnen and Sargent (2000) have suggested a procedure for developing an emulator which can be briefly described as follows:

1. *Determine the aim of the emulator*: any model should be developed for a specific goal. Usually we can identify four different goals: understanding the problem, predicting values of the response variable, performing optimization, and aiding the validation of the simulator.
2. *Identify the output variable*: the simulator usually has multiple response variables. However, current practice suggests that a separate metamodel should be developed for each output (single-response models).
3. *Identify the inputs and their characteristics*: determine the independent variables and specify the domain of applicability or experimental region X.
4. *Specify the required accuracy of the metamodel*: the range of accuracy depends on the goal of the metamodel. For example, if our main focus is on the optimization of a system, we cannot require a high level of complexity, since the optimization algorithms need simple models.

5. *Specify the metamodel*: in this step we should select a particular type of model among those proposed in the literature, e.g., polynomial regression models, Fourier metamodels, splines, neural networks, Kriging...
6. *Specify the type of experiment*: the experimental design problem is the selection of inputs for the computer code. The following main approaches are used to select designs (for more details see Sect. 4.2.4):

   • If we believe that interesting features of the simulator can be found in several parts of the experimental region, i.e., that no privileged areas exist, it seems natural to use designs that spread the points "evenly" to cover the full range of the input space. This allows the researcher to gather information about the relationship between the inputs and outputs for all regions of the domain of applicability. Furthermore, by covering the full range of input domain, space-filling designs could lead to good predictions over the entire input space, which is typically a fundamental goal in computer experiments. There are a number of ways to define what it means to spread points, and these lead to a variety of designs, usually called *space-filling* designs.
   • If we choose a particular type of metamodel, it is possible to formulate specific criteria for choosing a design and thus adopt an *optimal design*. Several criteria have been proposed in the literature. A recent class of design procedures which use a criterion based on the Fisher information matrix can be found in Zagoraiou and Baldi Antognini (2008).

7. *Fit the metamodel*: we run the simulation to obtain the output data from the inputs specified in step 6 for fitting the metamodel. From these data, we estimate the parameters of the metamodel.
8. *Validation of the fitted model*: starting from a test set of inputs we run simulations to validate the emulator, verifying the accuracy of the prediction (see Sect. 4.2.3). In the case of time-consuming and/or computationally intensive experiments, the researcher can apply "cheap" methods, such as cross-validation.

4.2.2 A Special Type of Emulator: The Kriging Technique

A wide variety of different modeling techniques are available in the literature. In this section, we mention some of them, and we focus our attention on the Kriging methodology, originally developed for geological applications but now a very important tool for producing accurate predictions of the output of a deterministic computer code.

The selection of a metamodel to approximate the true model as accurately as possible is a crucial problem. Generally, most of the metamodels in the literature are linear combinations of basis functions defined for an experimental region, and the unknown coefficients of the combination have to be estimated. Thus, when the simulator is deterministic, the construction of a surrogate can be viewed as an interpolation problem. Following this point of view, most metamodels take the form:

$$g(\mathbf{x}) = \sum_j B_j(\mathbf{x})\beta_j \qquad (4.1)$$

where B_j is a set of basis functions, β_j are unknown coefficients, and $\mathbf{x}$ is the input variable. If the basis function is polynomial, we obtain polynomial surrogates, which are widely used to model computer experiments. Often, the behavior of the data cannot be explained by the polynomial models. A solution to this problem is to use so-called splines, where the polynomials are defined in a piecewise way, i.e., several low-order polynomials are fit to the data, each in a separate range defined by the knots. Splines and polynomial models make the assumption that the data can be locally or globally fit by a polynomial. An alternative method which allows the data to be fitted in a less constrained form is known as the neural network technique; for more details, see Fang et al. (2006). Other bases for the construction of a metamodel have also appeared in the literature. For example, the Fourier basis can be used to approximate periodic functions.

Metamodels set up on a polynomial basis, spline basis or Fourier basis are competitive if the number of input variables is small, but extending them to a high-dimensional multivariate input can be difficult. Therefore, other techniques have been proposed. One of the most popular is the Kriging methodology. This kind of approach was originally proposed by a South African geologist, D.G. Krige (1951), for the analysis of geostatistical data. His work was later developed by Matheron (1962) and has become very popular in several applied contexts of the spatial statistics literature (see Ripley 1981 and Cressie 1993). Recently, this modeling approach has been widely used for the design and analysis of computer experiments (see for instance Sacks et al. 1989, Welch et al. 1992 and Bursztyn and Steinberg 2006).

Following Sacks et al. (1989), the Kriging approach consists of treating the deterministic response $y(\mathbf{x})$, i.e., the output of the simulator, as a realization of a stochastic process (random field) $Y(\mathbf{x})$ such that

$$Y(\mathbf{x}) = \mu(\mathbf{x}) + Z(\mathbf{x}) , \qquad (4.2)$$

where $\mu(\mathbf{x})$ denotes the global trend and $Z(\mathbf{x})$ represents the departure of the response variable $Y(\mathbf{x})$ from the trend. More precisely, $Z(\mathbf{x})$ is usually assumed to be a Gaussian stationary process with $E(Z(\mathbf{x})) = 0$, a constant variance σ_Z^2, and a non-negative correlation function between two inputs $\mathbf{x}$ and $\mathbf{w} \in X$.

$$\mathrm{Corr}[Z(\mathbf{x}), Z(\mathbf{w})] = R(\mathbf{x}, \mathbf{w}) , \qquad (4.3)$$

depending only on the displacement vector between any pair of points in X, and tending to 1 as the displacement vector goes to 0.

The correlation function should reflect the characteristics of the output. For a smooth response, a correlation function with some derivatives would be preferred, while an irregular response might call for a function with no derivatives. It is customary for the correlation to have the following property:

$$R(\mathbf{x}, \mathbf{w}) = \prod_j R_j(|x_j - w_j|), \tag{4.4}$$

i.e., products of one-dimensional correlations. Of special interest are those within the *power exponential family*:

$$R(\mathbf{x}, \mathbf{w}) = \prod_j \exp(-\theta_j |x_j - w_j|^p) \tag{4.5}$$

where $0 < p \leq 2$. We can also permit p to vary with j. The case $p = 1$ is the product of Ornstein–Uhlenbeck processes; they are continuous but not very smooth. A special type is the one-parameter exponential correlation function given by:

$$R(x, w) = \exp(-\theta |x - w|). \tag{4.6}$$

When $p = 2$ we have the so-called Gaussian correlation function, which is suitable for smooth and infinitely differentiable responses. An alternative choice for $R(x, w)$ is the linear correlation function:

$$R(x, w) = -\theta |x - w|. \tag{4.7}$$

Two different types of Kriging metamodels have been proposed in the literature depending on the functional form of the trend component:

- **Ordinary Kriging:** the trend is constant $\mu(\mathbf{x}) = \mu$ but unknown.
- **Universal Kriging:** the trend component depends on $\mathbf{x}$ and is modeled in a regressive way

$$\mu(\mathbf{x}) = \sum_{j=1}^{p} f_j(\mathbf{x})\beta_j = \mathbf{f}^t(\mathbf{x})\beta \tag{4.8}$$

where $f_1(\cdot), \ldots, f_p(\cdot)$ are known functions and $\beta = (\beta_1, \ldots, \beta_p)^t$ is the vector of the unknown parameters.

Kriging modeling lends itself to a sound theoretical methodology for the prediction of the output. Let

$$Y(\mathbf{x}) = \sum_{j=1}^{p} f_j(\mathbf{x}) \cdot \beta_j + Z(\mathbf{x}) = \mathbf{f}^t(\mathbf{x}) \cdot \beta + Z(\mathbf{x}) \tag{4.9}$$

be the model under study, $\mathbf{x}_1, \ldots, \mathbf{x}_n$ be the inputs,

$$\mathbf{Y}^n = (Y_1, \ldots, Y_n)^t = (y(\mathbf{x}_1), \ldots, y(\mathbf{x}_n))^t \tag{4.10}$$

be the set of outputs variables, and $y(\mathbf{x}_0) = Y_0$ be the output to predict. Furthermore, let $\hat{Y}_0$ denote a generic predictor of Y_0. Then the model is denoted

$$\begin{pmatrix} Y_0 \\ \mathbf{Y}^n \end{pmatrix} \sim \left[\begin{pmatrix} \mathbf{f}_0^t \\ \mathbf{F} \end{pmatrix} \beta, \sigma_Z^2 \begin{pmatrix} 1 & \mathbf{r}_0^t \\ \mathbf{r}_0 & \mathbf{R} \end{pmatrix} \right] \tag{4.11}$$

where $\mathbf{f}_0 = \mathbf{f}(\mathbf{x}_0)$ is the $(p \times 1)$ vector of regression functions for Y_0,

$$\mathbf{F} = (f_j(\mathbf{x}_i)) = \begin{pmatrix} \mathbf{f}_1(\mathbf{x}_1) & \cdots & \mathbf{f}_p(\mathbf{x}_p) \\ \vdots & & \vdots \\ \mathbf{f}_1(\mathbf{x}_n) & \cdots & \mathbf{f}_p(\mathbf{x}_n) \end{pmatrix} \tag{4.12}$$

is the $(n \times p)$ matrix of regression functions for the observed data, $\mathbf{r}_0 = (\mathbf{R}(\mathbf{x}_0 - \mathbf{x}_1),$
$\ldots, \mathbf{R}(\mathbf{x}_0 - \mathbf{x}_n))^t$ is the $(n \times 1)$ vector of correlations of $\mathbf{Y}^n$ and Y_0,

$$\mathbf{R} = \mathbf{R}_{i,j} = (R(\mathbf{x}_i, \mathbf{x}_j)) \tag{4.13}$$

is the $(n \times n)$ matrix of correlations among the entries of $\mathbf{Y}^n$, and β and σ_Z^2 are the unknown parameters.

If the correlation function is known, which is hardly ever the case in practical situations, the BLUP (best linear unbiased predictor) of Y_0 is given by

$$\hat{Y}_0 \equiv \mathbf{f}_0^t \hat{\beta} + \mathbf{r}_0^t \mathbf{R}^{-1} \left(\mathbf{Y}^n - \mathbf{F}\hat{\beta} \right) \tag{4.14}$$

where

$$\hat{\beta} = \left(\mathbf{F}^t \mathbf{R}^{-1} \mathbf{F} \right)^{-1} \mathbf{F}^t \mathbf{R}^{-1} \mathbf{Y}^n \tag{4.15}$$

is the generalized least squares estimator of β. Note that $\hat{Y}_0$ interpolates the data $(\mathbf{x}_i, y(\mathbf{x}_i))$ for $1 \leq i \leq n$, and that $\hat{Y}_0$ is a linear unbiased predictor of $Y(\mathbf{x}_0)$.

On the other hand, if the correlation function is unknown, $R(\cdot)$ can be written as

$$R(\cdot) = R(\cdot | \psi) \tag{4.16}$$

where ψ is an unknown parameter vector. Then it can be estimated by maximum likelihood, restricted maximum likelihood, cross-validation, or the posterior mode. In this case the predictor $\hat{Y}_0$ is termed the empirical best linear unbiased predictor, given by:

$$\hat{Y}_0 \equiv \mathbf{f}_0^t \hat{\beta} + \hat{\mathbf{r}}_0^t \hat{\mathbf{R}}^{-1} \left(\mathbf{Y}^n - \mathbf{F}\hat{\beta} \right) . \tag{4.17}$$

However, predictions are no longer linear. For a thorough reading, see Santner et al. (2003).

4.2.3 Accuracy of the Predictor

To validate an emulator, we must know the accuracy required of it, and this depends on its purpose. However, as is well known, in many cases there will be a trade-off between accuracy and complexity. Ideally, we want the departure of the emulator from the deterministic simulation code to be as small as possible over the experimental region, i.e.,

$$(\hat{y}(\mathbf{x}), y(\mathbf{x}))_L \rightarrow 0 \quad \text{for all } \mathbf{x} \in X , \tag{4.18}$$

where L is some measure of distance. Usually, the mean square error (MSE) of prediction at an untried point $\mathbf{x}$ is used to provide a measure of the overall model accuracy:

$$\mathrm{MSE}(\hat{y}(\mathbf{x})) = E\left(\hat{y}(\mathbf{x}) - y(\mathbf{x})\right)^2 . \tag{4.19}$$

Thus, one would need to know the values of the true model over the whole X. Therefore, the key element is to compare the predictions with the known true values in a test set. Possible ways to choose this kind of set are:

- In the case of "cheap" experiments, we may collect a large number of points, say N points $\mathbf{x}_i^*$, $i = 1, \ldots, N$ and calculate the empirical integrated mean squared error (EIMSE):

$$\mathrm{EIMSE} = \frac{1}{N} \sum_{i=1}^{N} \left(\hat{y}(\mathbf{x}_i^*) - y(\mathbf{x}_i^*)\right)^2 \tag{4.20}$$

  or the empirical maximum mean squared error (EMMSE)

$$\mathrm{EMMSE} = \max_{\mathbf{x}_i^*} \left(\hat{y}(\mathbf{x}_i^*) - y(\mathbf{x}_i^*)\right)^2 \tag{4.21}$$

  to check whether the metamodel satisfies the required accuracy.
- In general, since computer experiments are time-consuming and computationally intensive, the accuracy of the prediction will be assessed using adequate methods such as cross-validation. This method is based on iteratively partitioning the full set of available data into training and test subsets. For each partition, the researcher can estimate the model via the training subset and evaluate its accuracy using the test subset (Fang et al. 2006).

4.2.4 Experiments for a Kriging model

The experimental design problem for computer experiments is to select inputs for a computer code. The following main approaches are used in the literature to select designs: space-filling designs, and designs based on some optimality criterion.

4.2.4.1 Space-Filling Designs

As is recognized by many researchers, when no details on the functional behavior of the response variable are available, it is important to be able to obtain information from the entire design space. Therefore, design points should fill up the entire region in a "uniform fashion."

There are several statistical strategies that one might adopt to fill a given experimental region. One possibility is to select a simple random sample of points from the experimental region. However, other sampling schemes such as *stratified random sampling* are preferable, since *simple random sampling* is not completely satisfactory; e.g., with small samples and in high-dimensional experimental regions, it may present some clustering and fail to provide points in several portions of the domain.

Another method of generating designs that spread observations over the range of each input variable is so-called *latin hypercube sampling* (LHS). Introduced by McKay et al. (1979), LHS design is one of the most commonly used space-filling designs. An LHS design yielding n design points involves stratifying the experimental space into n equal probability intervals for each dimension, randomly selecting a value in each stratum, and then combining them in order to obtain a design point.

The "space-filling" property has inspired many statisticians to develop related designs. One class of designs maximizes the minimum Euclidean distance between any two points in the multidimensional experimental area. Other designs minimize the maximum distance or are based on the principle of comparing the distribution of the points with the uniform distribution (see Johnson et al. 1990, Koehler and Owen 1996 and also Santner et al. 2003).

4.2.4.2 Optimal Designs

An alternative way of choosing a design is to base the decision on some statistical criterion. After the type of metamodel has been chosen, the researcher can choose the design according to one of the many criteria proposed in the literature. The search for optimum designs for random field models is a rich but difficult research area. A good design for a computer experiment should facilitate accurate prediction, and most of the progress in this area has been made with respect to criteria based on functionals of the MSE, i.e.,

- The integrated mean squared error (IMSE) criterion:

$$\text{IMSE} = \int_X E\left[\hat{y}(\mathbf{x}) - y(\mathbf{x})\right]^2 d\mathbf{x} \qquad (4.22)$$

where the expectation is taken with respect to the random field.
- The maximum mean squared error (MMSE) criterion:

$$\text{MMSE} = \max_{\mathbf{x} \in X} E\left[\hat{y}(\mathbf{x}) - y(\mathbf{x})\right]^2 . \qquad (4.23)$$

Then, a design is said to be IMSE-optimal or MMSE-optimal if it minimizes (4.22) or (4.23), respectively. Observe that the IMSE and MMSE criteria can be considered generalizations of the A- and G-optimality (Silvey 1980), respectively. Both criteria can be calculated if the correlation function is known, which is impossible in practical situations. One possible way of overcoming this problem consists of starting with a space-filling design, e.g., a latin hypercube sampling scheme, in order to estimate the unknown correlation parameters, and then determining the IMSE-optimal or MMSE-optimal design, treating the obtained estimates as the true parameter values.

Lindley (1956) proposed the use of the change in entropy before and after collecting data as a measure of the information provided by an experiment. The entropy criterion for random fields is thus:

$$E(\Delta H(Y)) \tag{4.24}$$

where $\Delta H(Y)$ is the reduction in entropy after observing Y. A good design should maximize the expected reduction in entropy (see Shewry and Wynn 1987 and Bates et al. 1996).

An alternative criterion proposed in the literature is the maximum prediction variance. In this case, we choose to take observations where Kriging predictions are most uncertain, i.e., we select the design point which maximizes $\text{Var}(\hat{y}(\mathbf{x}))$.

A novel method based on the theory of optimal designs has been proposed by Zagoraiou and Baldi Antognini (2008). The main aim of the authors is to derive optimal designs for maximum likelihood estimation for ordinary Kriging with exponential correlation structure (4.6) using a criterion based on the Fisher information matrix. When the interest is mainly in the estimation of the trend parameter, they prove that the equispaced design is optimal for any sample size, while an optimal design for the estimation of the correlation parameter does not exist. Furthermore, the optimal strategy for the trend conflicts with the one for θ, since the equispaced design is the worst solution for estimating the correlation parameter. Hence, when the inferential purpose concerns the estimation of both the unknown parameters, the authors propose *geometric progression design*, namely a flexible class of procedures that allow the experimenter to choose a suitable compromise for the estimation precision of the two unknown parameters that at the same time guarantees a high efficiency for both.

4.3 Sequential Experiments for Kriging

4.3.1 Design and Analysis of Sequential Experiments

When experiments are carried out sequentially by the observer, at each step all of the information gathered up to that point is available in order to decide whether to observe any further and, if so, how to perform the next observations. Thus, the experimental decisions concerning the data collection process can evolve in an adaptive way on the basis of the experiment itself. *Sequential designs* are a type of experimental plan which consists of

1. A rule specifying at each stage the set of experimental conditions under which the next statistical unit(s) must be observed, and
2. A stopping rule, usually governed by economic constraints.

These rules are defined a priori in terms of the information gathered up to each given step, namely the data observed and the way in which the experiment itself has evolved.

In point of fact, there are many potential reasons for conducting the trial sequentially, in particular crucial economic demands (time–cost savings); the ability to exploit newly acquired data during the trial may allow one to pursue continuous quality improvement, which also results in a significant cost reduction.

There is another important factor in experiments in various fields of application, namely randomization. This is a rather loose term that has been brought into general use by R.A. Fisher; it is the action of assigning some features of the experiment by "controlled chance:" it may refer to the way in which experimental units are chosen, the order in which treatments are allocated to the units, and so on. It is nowadays commonly believed that a component of randomization in the design of the experiment is always required in order to protect against various types of bias (for instance accidental bias, selection bias, chronological bias, etc.), and it is also a fundamental tool for correct inferential procedures. However, there may be other reasons for introducing a probabilistic component into the experimental design that have not been sufficiently stressed in the literature, which result in an improvement in the convergence properties of the experiment itself, as in Baldi Antognini and Giovagnoli (2005).

In general, the choice of the design depends on several aspects which reflect the experimental aims: accurate inference about unknown parameters, accurate predictions at untried sites, cost savings, etc... Often these objectives can be defined as an optimization problem, but the solutions, the so-called *optimal design* or *target*, may depend on the unknown parameters (*local optimality*). This is typically the case in nonlinear problems, namely when the statistical model is nonlinear in the parameters of interest or the inferential aim consists of estimating a nonlinear function of the unknown parameters. In this context, sequential designs are essential, since they may represent a natural solution to the local optimality problem. In fact, by adopting a suitable response-adaptive procedure, the available outcomes can be used at each stage to estimate the unknown parameters, and thus the optimal design too; therefore, the allocations can be redirected in order to converge to the unknown target (*asymptotic optimality*). An example is the maximum likelihood design, which is a sequential randomized procedure based on the step-by-step updating of the optimal target by ML estimates (see for instance Baldi Antognini and Giovagnoli 2005).

However, the adoption of a particular sequential procedure may pose problems in relation to the correct inferential paradigm (Silvey 1980; Ford et al. 1985; Rosenberger et al. 1997; Giovagnoli 1999; Hu and Rosenberger 2003; Baldi Antognini and Giovagnoli 2006). If the design does not depend on the observed data, then it can be regarded as an ancillary statistic, and we can infer conditionally from the observed sequence of design points. In contrast, for all response-adaptive procedures, the design variability must be accounted for and we should argue unconditionally (Rosenberger and Lachin 2002). Thus, sequential experiments may be difficult to analyze, because the inferential methodology depends on the adopted design (Ford et al. 1985; Chaudhuri and Mykland 1993; Thompson and Seber 1996; Melfi and Page 2000; Aickin 2001). Recently, Baldi Antognini and Giovagnoli (2005) have shown that, under suitable conditions related to the statistical model and the allocation rule, conditional and unconditional inference tend to be the same asymptotically. This result can be applied when it is possible to simulate extensively from the computer code, but may be inappropriate for small samples, and/or when the simulation runs are very time/cost-expensive.

4.3.2 Recent Developments in Sequential Computer Experiments for Kriging with Applications

A vast amount of literature on sequential designs for computer experiments based on the Kriging methodology is available (Williams et al. 2000; Van Beers and Kleijnen 2003; Kleijnen and Van Beers 2004; Gupta et al. 2006; Romano 2006). In general, these methods start with a pilot study, where a small number n_0 of observations from a space-filling design will be gathered in order to get an initial estimation of the unknown parameters of the model; then, the sequential procedure is activated through the adoption of a suitable allocation rule, which specifies at each step the choice of the next design point; popular criteria are based on IMSE, MMSE, entropy, variance of prediction, etc... (see Sect. 4.2 for details). Finally, this procedure is stopped according to the specifications of the stopping rule, which is usually related to cost–time constraints or to the required inferential accuracy (i.e., when the improvement in precision is negligible).

For instance, the sequential Kriging design (SK) is one of the first proposals in this area (see Chap. 8). This is a deterministic response-adaptive procedure which selects at each step the design point at which the estimated variance of prediction is maximal. The SK procedure is very simple to implement, and at each step is sequential over the entire design region; however, it tends to get locked in the so-called *interesting areas*, namely in certain subsets of the experimental domain where the function assumes critical behavior (i.e., local or global maxima/minima).

4.3.3 "Application-Driven Sequential Designs"

This is the name of a sequential procedure recently introduced by Kleijnen and Van Beers (2004) for fitting the ordinary Kriging metamodel with one-parameter exponential correlation (4.6) or the linear correlation function in (4.7). Application-driven sequential designs (ADSD) can be briefly described as follows:

- The procedure starts with a pilot experiment generated by a LHS or a *maximin* design, which includes the extremes of the design space (since extrapolations via the Kriging methodology are not recommended); based on the simulations at these points, the correlation parameter is estimated by the cross-validation method.

- Using a space-filling approach, a given set of candidate points is selected (the authors suggest that the specific candidates halfway between the design points of the pilot study should be used), and at each candidate the variance of the predicted output is estimated by a cross-validation combined with jackknife technique.
- At each stage, the next design point (i.e., the winner among the candidate points for the actual simulation) is the one with maximum jackknife variance.
- The sequential procedure is stopped when no substantial improvement in terms of jackknife variance is observed.

The authors test the properties of the ADSD through two applications, namely hyperbolic and four-degree polynomial input–output functions, for small samples (i.e., $n_0 = 4$ and a total sample size N of up to 36). Assuming that the prediction errors induced by the procedure can be measured by EIMSE (4.20) and EMMSE (4.21), the authors show that ADSD gives better results than LHS with a prefixed sample of the same size.

4.3.4 A Modified Version via Randomization

Alessandro Baldi Antognini together with Paola Pedone have generalized the ADSD (Kleijnen and Van Beers 2004) by introducing the randomized sequential Kriging (RSK) design. This is a randomized response-adaptive procedure for fitting the universal Kriging metamodel with power exponential correlation (4.5), which can be described as follows:

- The experiment starts with a space-filled design (LHS), which includes the extremes of the design space and, based on the simulations on these points, the unknown parameters are estimated by the maximum likelihood method.
- Again, a given set of candidate points is selected using a space-filling approach, and at each candidate the variance of the predicted output is estimated.
- The choice of the next design point is made through a random assignment where the probability of selecting each candidate is proportional to the estimated prediction variance at this site obtained at the previous step.
- This sequential procedure is stopped when no substantial improvement in terms of EIMSE or EMMSE is observed.

With respect to ADSD, the particular features of RSK are that: (a) the assumed model is more general than the ordinary Kriging with one-parameter exponential correlation (4.6), and it is also more suitable in the case of smooth functions; (b) at each stage the parameters will be estimated by the ML method; (c) the choice of the design points is randomized: this random mechanism in the assignments is suggested in order to explore *all* of the design region without being trapped by local maxima/minima.

Table 4.1 Hyperbolic with $N = 19$

	RSK	ADSD	SK	LHS
EIMSE	4×10^{-6}	9×10^{-4}	0.008	0.006
EMMSE	4×10^{-5}	0.08	0.35	0.36

Table 4.2 Hyperbolic with $N = 36$

	RSK	ADSD	SK	LHS
EIMSE	10^{-7}	10^{-4}	8×10^{-4}	3×10^{-4}
EMMSE	10^{-6}	0.03	0.15	0.08

Table 4.3 4^0 polynomial with $N = 18$

	RSK	ADSD	SK	LHS
EIMSE	6×10^{-5}	0.17	0.58	0.59
EMMSE	9×10^{-4}	1.05	0.67	3.30

Table 4.4 4^0 polynomial with $N = 24$

	RSK	ADSD	SK	LHS
EIMSE	5×10^{-6}	0.01	0.27	0.25
EMMSE	6×10^{-5}	0.25	0.51	2.12

In Tables 4.1–4.4, we have compared the performances of RSK, ADSD, SK and LHS in terms of EIMSE and EMMSE with the same input–output functions considered in Kleijnen and Van Beers (2004).

The tables show that the introduction of a randomization component in the assignments makes it possible to overcome one well-known drawback of sequential procedures for Kriging, namely that the algorithms tend to linger for a long time around some critical areas. Furthermore, the choice between ordinary Kriging and universal Kriging seems to be almost irrelevant (Sasena 2002), whereas the random mechanism in the choice of the next design point combined with the maximum likelihood estimations of the unknown parameters is crucial. Observe that the RSK design in this case has performed better than the other procedures: the precision is high even with a small number of observations, and the efficiency of ADSD is reached by this design with approximately 40% fewer observations.

4.4 Robust Parameter and Tolerance Design via Computer Experiments

4.4.1 Robust Parameter and Tolerance Design: Crossed Arrays and Combined Arrays

Robust parameter design, introduced in the 1980s by the Japanese engineer G. Taguchi (Taguchi and Wu 1980), is an experiment-based statistical methodology that aims to find the nominal settings of design variables (parameters) that achieve a desired compromise between two objectives: optimizing the performance of a system by keeping the mean system performance around an ideal quality target and the variation around the mean to a minimum. This variation is caused by random variables (noise) related either to the external conditions (temperature, humidity, electromagnetic field, mechanical vibrations, way in which it is used) under which the product/process is operating, or to random variations in parameters due to manufacturing errors. Thus, random noise plays a major role in robust design.

When the design scope is extended to the specification of allowable deviations of parameters from nominal settings (tolerances), an *integrated parameter and tolerance design* problem arises. Here the additional objective of minimizing the production costs needed to fulfil tolerance specs will compete with the minimum variance objective.

The statistical methodology underlying robust design that is now widely accepted is dual response surface methodology, which estimates two surfaces, one for the mean and one for the variance (or the log-variance) of the process; see for instance Myers et al. (1992) and Myers and Montgomery (2002).

Let the system performance be described by a variable which depends on a set of controllable factors and a set of random noise factors: we write $Y(\mathbf{x}, \mathbf{Z})$ where the vector $\mathbf{x}$ stands for the levels of the controls and the vector $\mathbf{Z}$ for those of the noise. To explore the dependence of Y on $\mathbf{x}$ and $\mathbf{Z}$, one option is to run a *crossed-array* (Taguchi-type) experiment: some level combinations of the control factors (*inner array*) are chosen and tested across some level combinations of the random factors (*outer array*), which in the experimental setting are selected by the experimenters and are therefore regarded as fixed (nonrandom).

An alternative is a *combined array* experiment (see Vining and Myers 1990, Welch et al. 1990) with fixed levels of both the control and the noise variables. In such experiments, the in-process response is described by the model:

$$Y(\mathbf{x}, \mathbf{Z}) = \beta_0 + \beta^t \mathbf{x} + \mathbf{x}^t \mathbf{B} \mathbf{x} + \gamma^t \mathbf{Z} + \mathbf{x}^t \Delta \mathbf{x} + \varepsilon, \qquad (4.25)$$

where $\mathbf{Z}$ is the random noise vector, ε's are i.i.d. $N(0, \sigma^2)$ random errors, and ε and $\mathbf{Z}$ are independent. The constant β_0, the vectors β, γ and the matrices $\mathbf{B}$ and Δ consist of unknown parameters, and σ^2 is also usually unknown. It is also assumed that $E(\mathbf{Z}) = \mathbf{0}$ and that $\text{Cov}(\mathbf{Z})$ is known. After model (4.25) is fitted to the data, two response surfaces, one for the mean of Y as a function of the control factors $\mathbf{x}$, and one for the variance of Y, also in terms of the control factors, are obtained analytically from (4.25) by taking expectations.

Both approaches (crossed array and combined array) have some drawbacks: we now briefly describe the method suggested in Giovagnoli and Romano 2008 which generalizes both approaches making use of a stochastic simulator.

4.4.2 The Proposed Simulation Protocol and Its Application to the Integrated Design of Parameters and Tolerances

Let us divide the random factors into two independent vectors, $\mathbf{Z}_1$ and $\mathbf{Z}_2$. The random vector $\mathbf{Z}_1$ will include the variables that we are going to simulate stochastically, while the remaining set of noise factors $\mathbf{Z}_2$ will be given fixed levels $\mathbf{z}_2$ for some different choices of $\mathbf{z}_2$. At the same time, different levels $\mathbf{x}$ of the control factors are also chosen for the experimentation. The computer experiment is performed by stochastically simulating the noise $\mathbf{Z}_1$ for chosen pairs $(\mathbf{x}, \mathbf{z}_2)$, and the sample mean and variance of the observed responses are calculated. The hypothesis underlying the proposed approach is that a simulation code of the physical system is available and that noise factors are explicitly modeled in the code. The normality of $\mathbf{Z}_1$ can often be reasonably assumed. Clearly this method requires that existing models are modified to take into account the additional variability introduced by simulating the randomness of $\mathbf{Z}_1$.

In the case $\mathbf{Z}_1 = \emptyset$ (or, equivalently, $\mathbf{Z}_2 = \mathbf{Z}$), the simulation becomes nonstochastic and the model reduces to model (4.25) for the combined array approach, but without the experimental error. In the case $\mathbf{Z}_1 = \mathbf{Z}$ (or, equivalently, $\mathbf{Z}_2 = \emptyset$), one gets the crossed-array approach with independent and normally distributed noises.

This method can be extended to a several-stage procedure if noise factors are simulated by adding them sequentially, one after another, to $\mathbf{Z}_1$. This allows sequential assessment of the way in which each single noise factor affects the total variability.

Other proposals for conducting robust design studies (see Lehman et al. 2004, Bates et al. 2006) have recently been made in the context of computer experiments, but they do not envisage any stochastic simulation.

Sometimes robust design aims to set both design parameters and tolerance specifications. External noise variables are totally out of the designer's control, but internal noise variables, which represent random deviations in design parameters due to part-to-part variations induced by uncontrollable manufacturing errors, are partially controllable. In this case, the size of the common variability can be decided upon by the designer by choosing components of suitable quality. It seems natural, at the experimental stage, to take this variability as a control factor and simulate the internal noise accordingly at each level of this factor. Hence, internal noise variables, under the common assumption of independence and a normal distribution with zero mean, are natural candidates for inclusion in the $\mathbf{Z}_1$ vector of the proposed method.

After the modified dual response approach is applied, the in-process mean and variance of the response are obtained as functions of the control factors. These functions, coupled with cost functions which specify how individual tolerances affect production cost (the tighter the tolerances, the more expensive the manufacturing processes), are the ingredients for tackling the integrated parameter and tolerance design problem (see Li and Wu 1999; Romano et al. 2004) as an optimization problem under constraints.

4.4.3 An Application to Complex Measurement Systems

The new method was applied to the design of a high-precision optical profilometer (Fig. 4.1). This is a measuring device that inspects the surface of mechanical parts and then reconstructs the relevant profile. Its particular features are that the surface is inspected without contact and that measurement uncertainty is very low (in the range 10^1–10^2 nm).

An innovative prototype, based on the technique of white-light interferometry, has been recently realized at the Department of Mechanical Engineering of Cagliari University, Italy (see Baldi et al. 2002; Baldi and Pedone 2005). The measurement process combines the analog treatment of optical signals with the numeric processing of digitized images. The white light, emitted from a source, is separated by a polarized beam splitter into two beams that head to the mirror and the testpiece respectively. After reflection, the beams are recombined, producing interference fringes (alternating light and dark bands) that are collected by a video camera. The variation in light intensity at a given point in the image is represented as a function

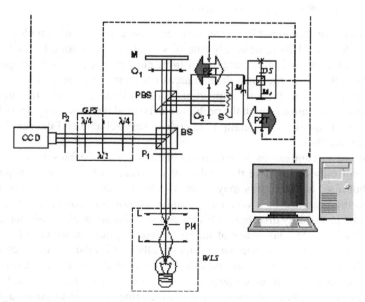

Fig. 4.1 Scheme of the optical profilometer

of the optical path difference; it theoretically corresponds to the sinc^2 function and peaks where the optical path difference equals zero, which realizes the condition of maximum positive interference. By processing images collected for a number of equispaced positions of the mirror, which is displaced along the x-axis by a piezo-electric transducer, a sample of points on the modulation curve at each pixel of the image is obtained. Estimating the abscissa of the maximum modulation for all pixels provides information on the surface profile. Relevant noise factors are the errors in the two coordinates of each sample point.

An experiment was performed at the Department of Mechanical Engineering of Cagliari University, Italy, with the objective of improving the measurement quality of the profilometer, i.e., minimizing bias and uncertainty in the profile reconstruction. As there are several sources of noise in the system, robust design was called for. Since experimenting on the prototype would have been exceedingly expensive, the whole measurement process was reproduced by a suitable simulator. The simulator incorporated both theoretical and experimental knowledge on the physical mechanisms involved. The measurement variance predicted by the simulator was also validated by comparing it with that obtained by replicated measurements of the prototype (Baldi et al. 2006).

4.4.4 Other Potential Applications

It is useful to clarify the conditions most suited to the utilization of the proposed method, and thus to identify applications which can benefit most from it. Some of the most congenial set-ups are now described.

- The simulator should not be exceedingly expensive in terms of completion times for single runs. This is generally true of stochastic simulators, but not just for them. Consider for instance the use of finite element codes to solve engineering design problems in the most diverse areas (mechanical components, buildings, electrical and electromagnetic apparatus, fluid-dynamic systems, etc.). In the not too infrequent case in which the problem can be modeled using a system of linear differential equations, the computing code is very reliable and the running time is relatively short.
- There are circumstances in which simulating noise variables may be convenient not only to improve the quality of the solution to a robust design problem but also, although this may seem paradoxical, to save costs. This may be due to the presence of a very large number of noise variables which all share the same probability distribution because of their common physical source. Consider for instance a simulator of microeconomic system in which a large number of individual—independent and identically distributed—choices come into play. Simulation from a common distribution is evidently more appropriate in this case, as well as more convenient. This is true of external noise factors, and even more so of internal ones, like manufacturing errors induced on a number

of similar part features, such as flat faces, bores, cylindrical surfaces, all machined by the same physical process (milling, drilling, turning). This way of simulating internal noise was applied in the case study mentioned above, on the optical profilometer. Here, the same piezoelectric transducer repeatedly moves a mirror by a nominally constant distance. Random errors in successive displacements, though independent, come from the same distribution. Under these circumstances, all of the independent identically distributed internal noise variables can be cumulatively accounted for by just one tolerance factor. This may yield a considerable saving in terms of runs. Suppose there are two control factors and ten internal noise variables of this kind. If we take two levels per control and three levels per noise variable and cross them fully, as is done in a Taguchi-type experiment, an outer array of $2^2 \times 3^{10} = 236,156$ runs will result. This number is far larger than the reasonable sampling size (800) obtained by running 100 simulations of ten tuples drawn from the common underlying distribution, for each combination of tolerance and control factor levels, say 2^3.

- Another interesting application field concerns a class of systems sometimes referred as *hybrid* systems, incorporating both hardware and software components. The hardware part collects or generates analog information which, in turn, after being converted into a digital format, is further processed by some software module. Such systems are already common in our daily life, and are becoming more so. Familiar examples include cellular phones and household appliances like dishwashers, food processors, etc. Sophisticated software processes the electrical signals to permit an acceptable quality of mobile communication, even in hostile external conditions. Household machines are provided with carefully designed digital controllers which can modify, in real time, the operating conditions of the machine in order to obtain optimal performance (effective operations, energy saving, noise reduction, vibration control and safety control). The design of hybrid systems is particularly challenging. Although the design techniques for the hardware and software components are highly heterogeneous, they must be perfectly coupled for effective design. The option here is to replace, at the design stage, the physical part of the system with a simulation model. In these circumstances, the whole system can be represented by a computer code, which is very easy to handle for design purposes. It is sensible to forecast that the software parts of hybrid systems will become dominant and more and more complex. This will make these systems easier to simulate (numerical treatment is a computer code) and even more fully designed (numerical treatment can deal with a huge number of design variables). This opens the way to intelligent and innovative products. The optical profilometer mentioned before is a hybrid system. Reconstruction of the microgeometry requires a massive numerical treatment of several images collected by a digital video camera after a number of costly optical components have processed the white light. Measurement and diagnosis systems are in fact one of the most interesting sectors where design via simulators is applicable. Allow us to mention three more cases: the process for the dimensional and geometric control of manufactured parts by coordinate measuring machines (CMM) (Romano and Vicario 2002a); the supervising systems for fault diagnosis in industrial processes (Romano and Kinnaert 2006); and the fixed-bed gasifier in a clean plant for obtaining energy from coal (Cocco et al. 2008). In all of these applications, a robust design approach has been applied to stochastic simulators (developed ad hoc) of the processes involved.

References

Aickin, M.: Randomization, balance, and the validity and efficiency of design-adaptive allocation methods. J. Stat. Plan. Infer. **94**, 97–119 (2001)

Alexandrov, N., Dennis, J.E. Jr., Lewis, R.M., Torczon, V.: A trust region framework for managing the use of approximation models in optimization. Struct. Optimiz. **15**(1), 16–23 (1998)

Baldi Antognini, A., Giovagnoli, A.: On the large sample optimality of sequential designs for comparing two or more treatments. Seq. Anal. **24**, 205–217 (2005)

Baldi Antognini, A., Giovagnoli, A.: On the asymptotic inference for response-adaptive experiments. Metron **64**, 29–45 (2006)

Baldi Antognini, A., Giovagnoli, A., Romano, D.: Implementing asymptotically optimal experiments for treatment comparison. Presented at the 5th ENBIS Conf., Newcastle, UK, Sept. (2005); available on CD-ROM

Baldi, A., Pedone, P.: Caratterizzazione numerico sperimentale dei rotatori geometrici nell'interferometria in luce bianca. Proc. of XXXIV AIAS Conference, Milan, Italy, pp. 203–204, 14–17 Sept. (2005); extended version on CD-ROM

Baldi, A., Giovagnoli, A., Romano, D.: Robust design of a high-precision optical profilometer by computer experiments. Presented at the 2nd ENBIS Conf., Rimini, 23–24 Sept. (2002)

Baldi, A., Ginesu, F., Pedone, P., Romano, D.: Performance comparison of white light-optical profilometers, Proc. 2006 SEM (Society for Experimental Mechanics) Annual Conf., St. Louis, MO, USA, paper no. 378 on CD-ROM, ISBN 0-912053-93-X, 4–7 June (2006)

Bashyam, S., Fu, M.C.: Optimization of (s, S) inventory systems with random lead times and a service level constraint, Manag. Sci. **44**(12), 243–256 (1998)

Bates, R.A., Buck, R.J., Riccomagno, E., Wynn, H.P.: Experimental design and observation for large systems. J. Roy. Stat. Soc. B **58**, 77–94 (1996)

Bates, R.A., Kenett, R.S., Steinberg, D.M., Wynn, H.P.: Achieving robust design from computer simulation. Qual. Technol. Quant. Manag. **3**(2), 161–177 (2006)

Bursztyn, D., Steinberg, D.M.: Comparisons of designs for computer experiments. J. Stat. Plan. Infer. **136**, 1103–1119 (2006)

Box, G.E.P., Wilson, K.B.: On the experimental attainment of optimum conditions. J. Roy. Stat. Soc. B **13**, 1–45 (1951)

Chaudhuri, P., Mykland, P.A.: Nonlinear experiments: optimal design and inference based on likelihood. J. Am. Stat. Assoc. **88**, 538–546 (1993)

Cressie, N.A.C.: Statistics for Spatial Data. Wiley, New York (1993)

Cocco, D., Romano, D., Serra, F.: Caratterizzazione di un gassificatore a letto fisso mediante la metodologia della pianificazione degli esperimenti. Proc. of 63° Congresso Associazione Termotecnica Italiana, Palermo, Italy, 23–26 Sept. (2008)

Fang, K.T., Li, R., Sudjianto, A.: Design and Modelling for Computer Experiments. Chapman and Hall/CRC, London (2006)

Farhang-Mehr, A., Azarm, S.: An information-theoretic performance metric for quality assessment of multi-objective optimization solution sets. J. Mech. D-T ASME **125**, 655–663 (2003)

Ford, I., Titterington, D.M., Wu, C.F.J.: Inference and sequential design. Biometrika **72**, 545–551 (1985)

Ghosh, B.K., Sen, P.K.: Handbook of Sequential Analysis. Marcel Dekker, New York (1991)

Giovagnoli, A.: Sul ruolo della probabilità e della verosimiglianza nella programmazione degli esperimenti. Statistica **59**, 559–580 (1999)

Giovagnoli, A., Romano, D.: Robust design via experiments on a stochastic simulator: a modified dual response surface approach. Qual. Reliab. Eng. Inter. **24**, 401–416 (2008)

Gupta, A., Ding, Y., Xu, L.: Optimal parameter selection for electronic packaging using sequential computer simulations. J. Manuf. Sci. Eng. T ASME **128**, 705–715 (2006)

Hu, F., Rosenberger, W.F.: Optimality, variability, power: evaluating response-adaptive randomization procedures for treatment comparisons. J. Am. Stat. Assoc. **98**, 671–678 (2003)

Johnson, M.E., Moore, L.M., Ylvisaker, D.: Minimax and maxmin distance design. J. Stat. Plan. Infer. **26**, 131–148 (1990)

Kleijnen, J.P.C., Sargent, R.G.: A methodology for fitting and validating metamodels in simulation. Eur. J. Oper. Res. **120**, 14–29 (2000)

Kleijnen, J.P.C., Van Beers, W.C.M.: Application-driven sequential designs for simulation experiments: Kriging metamodeling. J. Oper. Res. Soc. **55**, 876–883 (2004)

Koehler, J.R., Owen, A.B.: Computer experiments. In: Ghosh, S., Rao, C.R. (eds) Handbook of Statistics, vol. 13. Elsevier, Amsterdam, pp 261–308 (1996)

Lehman, J.S., Santner, T.J., Notz, W.I.: Designing computer experiments to determine robust control variables. Stat. Sinica **14**, 571–590 (2004)

Li, W., Wu, C.F.J.: An integrated method of parameter design and tolerance design. Qual. Eng. **11**, 417–425 (1999)

Lin, Y., Mistree, F., Allen, J.K.: A sequential exploratory experimental design method: development of appropriate empirical models in design. In: Chen, W. (ed.), ASME 30th Conf. of Design Automation, Salt Lake City, UT, USA, ASME Paper no. DETC2004/DAC-57527, 28 Sept.–2 Oct. (2004)

Lindley, D.V.: On a measure of information provided by an experiment. Ann. Math. Stat. **27**, 986–1005 (1956)

Matheron, G.: Traité de géostatistique appliquée, Vol. 1, Memoires du Bureau de Recherches Geologiques et Miniéres. Editions Bureau de Recherches Geologiques et Minieres (vol. 24), Paris (1962)

McKay, M.D., Beckman, R.J., Conover, W.J.: A comparison of three methods for selecting values of input variables in the analysis of output from a computer code. Technometrics **21**, 239–245 (1979)

Melfi, V., Page, C.: Estimation after adaptive allocation. J. Stat. Plan. Infer. **87**, 353–363 (2000)

Montgomery, D.C.: Design and Analysis of Experiments, 5th edn. Wiley, New York (2005)

Myers, R.H., Montgomery D.C.: Response Surface Methodology, 2nd edn. Wiley, New York (2002)

Myers, R.H., Khuri, A.I., Vining, G.G.: Response surface alternatives to the Taguchi robust parameter design approach. Am. Stat. **46**(2), 131–139 (1992)

Pedone, P., Vicario, G., Romano, D.: Kriging-based sequential inspection plans for coordinate measuring machines. ENBIS-DEINDE Conf., Turin, Italy, 11–13 April (2007); available on CD-ROM

Ripley, B.: Spatial Statistics. Wiley, New York (1981)

Romano, D.: Sequential experiments for technological applications: some examples. Proc. XLIII Scientific Meeting of the Italian Statistical Society—Invited Session: Adaptive Experiments, Turin, Italy, pp 391–402, 14–16 June (2006)

Romano, D., Kinnaert, M.: Robust design of fault detection and isolation systems. Qual. Reliab. Eng. Int. **22**(5), 527–538 (2006)

Romano, D., Varetto, M., Vicario, G.: A general framework for multiresponse robust design based on combined array. J. Qual. Technol. **36**(1), 27–37 (2004)

Romano, D., Vicario, G.: Inspecting geometric tolerances: uncertainty in position tolerances control on coordinate measuring machines. Stat. Meth. Appl. **11**(1), 83–94 (2002a)

Romano, D., Vicario, G.: Reliable estimation in computer experiments on finite element codes. Qual. Eng. **14**(2), 195–204 (2002b)

Rosenberger, W.F., Flournoy, N., Durham S.D.: Asymptotic normality of maximum likelihood estimators from multiparameter response-driven designs. J. Stat. Plan. Infer. **60**, 69–76 (1997)

Rosenberger, W.F., Lachin, J.M.: Randomization in Clinical Trials. Wiley, New York (2002)

Sacks, J., Welch, W., Mitchell, T.J., Wynn, H.P.: Design and analysis of computer experiments. Stat. Sci. **4**, 409–435 (1989)

Santner, J.T., Williams, B.J., Notz, W.J.: The Design and Analysis of Computer Experiments. Springer, New York (2003)

Sasena, M.J.: Flexibility and efficiency enhancements for constrained global design optimization with Kriging approximations. Ph.D. thesis, University of Michigan, MI (2002)

Shewry, M.C., Wynn, H.P.: Maximum entropy sampling. J. Appl. Stat. **14**, 165–170 (1987)

Silvey, S.D.: Optimal Designs. Chapman & Hall, London (1980)

Taguchi, G., Wu, Y.: Introduction to Off-line Quality Control. Central Japan Quality Control Association, Nagoya (available from American Supplier Institute, Romulus, MI) (1980)

Thompson, S.K., Seber, G.A.F.: Adaptive Sampling. Wiley, New York (1996)

Van Beers, W.C.M., Kleijnen, J.P.C.: Kriging for interpolation in random simulation. J. Oper. Res. Soc. **54**, 255–262 (2003)

Van Beers, W.C.M., Kleijnen, J.P.C.: Customized sequential designs for random simulation experiments: Kriging metamodeling and bootstrapping (Discussion Paper no. 55). Tilburg University, Holland (2005)

Vining, G.G., Myers, R.H.: Experimental design for estimating both mean and variance functions. J. Qual. Tech. **28**, 135–147 (1990)

Wang, G.G., Dong, Z., Aitchison, P.: Adaptive response surface method—a global optimization scheme for computation-intensive design problems. J. Eng. Opt. **33**(6), 707–734 (2001)

Welch, W.J., Buck, R.J., Sacks, J., Wynn, H.P., Mitchell, T.J., Morris, M.D.: Screening, predicting, and computer experiments. Technometrics **34**, 15–25 (1992)

Welch, W.J., Yu, T.-K., Kang, S.M., Sacks, J.: Computer experiments for quality control by parameter design. J. Qual. Tech. **22**, 15–22 (1990)

Williams, B.J., Santner, T.J., Notz, W.I.: Sequential design of computer experiments to minimize integrated response functions. Stat. Sinica **10**, 1133–1152 (2000)

Wujek, B.A., Renaud, J.E.: New adaptive move-limit management strategy for approximate optimization, Parts 1 and 2. AIAA J. **36**(10), 1911–1934 (1998)

Zagoraiou, M., Baldi Antognini, A.: On the optimal designs for Gaussian ordinary Kriging with exponential correlation structure. Appl. Stoch. Models Bus. Ind. (in press) (2008)

Part II
Technological Process Innovation

Chapter 5
Design for Computer Experiments: Comparing and Generating Designs in Kriging Models

Giovanni Pistone and Grazia Vicario

Abstract The selection of design points is mandatory when the goal is to study how the observed response varies upon changing the set of input variables. In physical experimentation, the researcher is asked to investigate a number of issues to gain valuable inferences. Design of experiments (D.o.E.) is a helpful tool for achieving this goal. Unfortunately, designing a computer experiment (CE), used as a surrogate for the physical one, differs in several aspects from designing a physical experiment. As suggested by the pioneers of CEs, the output can be predict by assuming Gaussian responses and that covariance depends parametrically on the distance between the locations, according to the Kriging model. Latin hypercube (LH) training sets are used in most cases. In this chapter, we discuss the influence of LHs on the prediction error of the conditional expectation step of the Kriging model using examples. Our suggestion is to perform these preliminary tests in order to assess which class of LH seems to fit the specific application.

5.1 Introduction

To represent the behavior of physical systems, it is common practice to resort to a mathematical model, such as:

$$y = f(x) \tag{5.1}$$

Giovanni Pistone
Politecnico di Torino, Department of Mathematics
Corso Duca degli Abruzzi, 24, 10129 Torino, Italy
e-mail: giovanni.pistone@polito.it

Grazia Vicario
Politecnico di Torino, Department of Mathematics
Corso Duca degli Abruzzi, 24, 10129 Torino, Italy
e-mail: grazia.vicario@polito.it

Pasquale Erto, *Statistics for Innovation*
ISBN 978-88-470-0814-4, © Springer 2009

where $x = (x_1, x_2, \ldots, x_s) \in \mathbf{D}$ is the vector of input variables (without distinguishing between *control variables* or *engineering variables*, *noise variables* and *model variables*), $\mathbf{y} \in \mathbb{R}$ is the output variable and $\mathbf{D}$, the input variable space, is a subset of $\mathbb{R}^s$ (Santner et al. 2003). Model (5.1) may imply the solution of ordinary or partial differential equations defined in a time–space domain, as in the case of the ubiquitous finite element method (Zienkiewicz 1971), the solution of a set of equations (either linear or not); furthermore, the function f may not have an analytic solution. It may be that the solution to such a system of equations is not available in closed form, or that the numerical computation of it takes an unacceptably long time to obtain, or that only an approximation of the true model (5.1) is available. Consequently, computer simulation may provide hints about the actual relationship between the output and input variables (see Fig. 5.1).

It is common practice for computer experiment (CE) practitioners to seek an approximate model:

$$y = g(x) \tag{5.2}$$

which is hopefully close enough to the real one in the domain of interest and lends itself more easily to numerical evaluation. Such an approximate model of an input/output function may be referred to as a *metamodel* (Kleijnen 1987) or, equivalently, an *emulator* (Sacks et al. 1989a,b). Such a metamodel may imply rather complex functions, not just low-order polynomials, a popular choice in response surface methodology (Box et al. 1987).

A metamodel is first estimated in a given class of models and subsequently it produces a prediction of interest on the basis of a set of data (*training set*). The choice of a data set is a problem classically termed "design of experiments" (D.o.E.). The appropriateness of a data set must be evaluated with respect to criteria that are strongly affected both by the metamodel chosen and the aim of the experiment. The applicability of the basic principles of classical D.o.E.—i.e., replication, randomization and blocking—is questioned in CE. Moreover, effects of uncontrolled factors and measurement uncertainties are the main sources of error in physical experiments. With CE, there are no measurement uncertainties, since repeated runs with the same input lead to exactly the same responses, and the effects of uncontrolled factors are present, possibly with an interpretation that is different from the physical case. Uncertainty is related to a lack of knowledge of the exact relationship between input variables and response (in D.o.E., this discrepancy is referred to as "model bias").

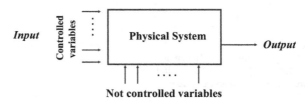

Fig. 5.1 Graphical representation of physical experimentation: a number of experiments are carried out according to changes in the input variables in order to acquire knowledge about the causal relationships between the input and output variables of a (usually quite complex) physical system

Therefore, the selection of an experimental design in CE is crucial to achieving an efficient and informative model. In principle, the problem is that of finding a good design, namely a set of experimental points $\mathbf{D}_n = \{\mathbf{x}_1, \mathbf{x}_2, \ldots, \mathbf{x}_n\} \subset \mathbb{R}^d$ for a given number of runs n; the CE approach involves uniform coverage of the experimental region (in which we want to predict the response y) and any number of levels for each factor.

In classical D.o.E., the minimization of some function of the covariance matrix containing the least squares estimates for the model parameters is an accepted principle, together with work aimed at ensuring the independence of estimates. It is suggested in the literature on CE that the principle should be to make the deviation between the true model (5.1) and the metamodel (5.2):

$$\mathrm{Dev}\,(\mathbf{x}_i; f, g) = f(\mathbf{x}_i) - g(\mathbf{z}_i) \tag{5.3}$$

as small as possible for any point $\mathbf{x}_i \in \mathbf{D}_n$, for $i = 1, 2, \ldots, n$, in the experimental region (there is no restriction on supposing that $\mathbf{D}_n$ is a d-dimensional cube, since all input variables may be rescaled; therefore the experimental region may be the unit cube C^d, without any loss of generality, where d is the number of the variables); see the discussion in Santner et al. (2003), Sasena et al. (2002), and Park et al. (2002)

It has been suggested by Sacks et al. (1989a,b) that the joint use of a *Kriging* model as a metamodel together with latin hypercube (LH) designs (Mckay et al., 1979; Loh, 1996) is the best option in CE when no specific model is imposed by the application itself. In the present chapter, we discuss the behaviors of different LH designs in minimizing the overall error of prediction for a regular grid. It has frequently been observed that some LH designs are not desirable because they do not ensure sufficient coverage of the design space. We hope that our results will be useful for clarifying the coverage criteria in quantitative terms.

5.2 Kriging Prediction

When the output of a CE is modeled according to the suggestion of J. Sacks, a *Kriging* model is used that considers the response $y(\mathbf{x})$, for $\mathbf{x} \in \mathbf{D} \subset \mathbb{R}^d$, to be a realization of a Gaussian random process $Y(\mathbf{x})$:

$$Y(x) = \beta' \mathbf{f}(\mathbf{x}) + Z(\mathbf{x}) \tag{5.4}$$

where $\mathbf{f}(\mathbf{x}) = (f_1(\mathbf{x}), f_2(\mathbf{x}), \ldots, f_m(\mathbf{x}))'$ is a set of specified trend functions, $\beta = (\beta_1, \beta_2, \ldots, \beta_m)'$ is a set of (usually unknown) parameters, and $Z(\mathbf{x})$ is a Gaussian random process with zero mean and stationary covariance over $\mathbf{D}$; therefore:

$$E[Y(\mathbf{x})] = \beta' \mathbf{f}(\mathbf{x})$$
$$\mathrm{Cov}\,(Y(\mathbf{x}), Y(\mathbf{x} + \mathbf{h})) = \sigma_Z^2 R(\mathbf{h}; \theta) \tag{5.5}$$

where σ_Z^2 is the process variance, and R is the stationary correlation function (SCF), which only depends on the displacement vector $\mathbf{h}$ between any pair of points in $\mathbf{D}$ and on the vector parameter θ.

Model (5.4) is known as "universal Kriging," and was created by a group of statisticians (Sachs et al. 1989a,b) at the end of the 1980s. Such a suggested framework for deterministic CEs appears to conflict with its origins: Kriging models are named after a South African engineer, Daniel G. Krige (1951), who first used them to predict noisy spatial responses from (generally) a small number of observations when analyzing mining data. The two contexts are quite different: the former is refers to phenomena affected by noise, and so the use of statistical models is valid; the latter assumes that the responses of a CE are the realization of a stochastic process, and makes much of the correlation structure dictated by the process.

The original suggestion, which is still the most popular choice for the correlation function, comes from the power exponential family:

$$R(h;\theta) = \prod_{l=1}^{d} \exp\{-\theta_l |h_l|^p\} = \exp\left\{-\sum_{l=1}^{d} \theta_l |h_l|^p\right\} \quad \text{with } 0 < p \le 2, \quad (5.6)$$

where $\theta = (\theta_1, \theta_2, \ldots, \theta_d, p)'$, θ_l are positive scale parameters and p is a common smoothing parameter. Parameter θ_l, $l = 1, 2, \ldots, d$, represents the rapidity with which the correlation decays in direction l with increasing distance h_l. The assumption in (5.6) that the positive correlation between outputs diminishes as the distance between input sites increases is the formalization of Krige's original idea.

Most practitioners (Santner et al. 2003) suggest that Bayesian estimators should be used to analyze the output of CE; prior information on the set $Y^n = (Y(\mathbf{x}_1), Y(\mathbf{x}_2), \ldots, Y(\mathbf{x}_n))'$ of process variables at $\mathbf{x}^n = (\mathbf{x}_1, \mathbf{x}_2, \ldots, \mathbf{x}_n)$ (also named the *training data*) is used to predict the unknown output $Y(\mathbf{x})$ at a new site $\mathbf{x}_0$. Then, under the hypothesis (consistent with (5.5) and (5.6)) that the joint random variable $(Y(\mathbf{x}_0), Y(\mathbf{x}_1), Y(\mathbf{x}_2), \ldots, Y(\mathbf{x}_n))$ is a multivariate normal, $N[(\mathbf{f}_0', \mathbf{F})'\beta, \sigma_Z^2 \Sigma]$, with $\Sigma = \begin{pmatrix} 1 & \mathbf{r}_0' \\ \mathbf{r}_0 & \mathbf{R} \end{pmatrix}$, the conditional mean of $Y(\mathbf{x})$ for the process data at the untried point $\mathbf{x}_0$, $\hat{Y}_0 = \mathrm{E}(Y(\mathbf{x}_0)|\mathbf{Y}^n)$ is:

$$\hat{Y}_0 = \mathbf{f}_0'\beta + \mathbf{r}_0' \mathbf{R}^{-1}(\mathbf{Y}^n - \mathbf{F}\beta) \quad (5.7)$$

where $\mathbf{f}_0$ is the $m \times 1$ vector of the trend functions in $\mathbf{x}_0$, $\mathbf{F}$ is the $n \times m$ matrix $\{f_j(\mathbf{x}_i)\}_{\substack{i=1,\ldots,n \\ j=1,\ldots,m}}$ of the trend functions computed in $(\mathbf{x}_1, \mathbf{x}_2, \ldots, \mathbf{x}_n)$, $\mathbf{r}_0$ is the correlation vector $(R(\mathbf{x}_0 - \mathbf{x}_1), \ldots, R(\mathbf{x}_0 - \mathbf{x}_n))'$, and $\mathbf{R}$ is the $n \times n$ correlation matrix whose (i, j) element is $R(\mathbf{h}_{ij} = \mathbf{x}_i - \mathbf{x}_j)$.

As the predictor $\hat{Y}_0 = \mathrm{E}(Y(\mathbf{x}_0)|\mathbf{Y}^n)$ minimizes the mean squared prediction error, $\mathrm{E}[(\hat{Y}_0 - Y_0)^2]$, it is the best linear unbiased predictor (BLUP) of $Y(\mathbf{x}_0)$, and it is also the unique one.

The mean squared prediction error

$$\mathrm{MSPE}\left[\hat{Y}_0\right] = \mathrm{E}\left[\left(\hat{Y}_0 - Y(\mathbf{x}_0)\right)^2\right] = \sigma_z^2\left(1 - \mathbf{r}_0' \mathbf{R}^{-1} \mathbf{r}_0\right) \quad (5.8)$$

usually called the *Kriging variance*, is a measure of prediction uncertainty. It is worth noting that it is large when $\mathbf{x}_0$ is far from the experimental points and small when it is close to them; it vanishes at the experimental points because of the interpolatory nature of Kriging.

Equations 5.7 and 5.8 hold only if β and $R(\mathbf{h};\theta)$ are known (rarely the case in practical situations), but we use them in a methodological context. We are going to compare the efficiencies of different training sets in toy examples for each value of β and $\mathbf{R}$; in other words, we do not consider the relative efficiencies of different training sets in terms of estimating unknown parameters.

5.3 A Class of Designs: Fractions of a Regular Grid

The correlation model (5.6) is defined for $\theta_l > 0$, $l = 1, 2, \ldots, d$, and p between 0 and 2. It is known that these conditions are necessary and sufficient for a function of that form to be positive definite in Euclidean space and, therefore, for a stationary Gaussian process to exist with that covariance function.

In the present computations we are going to assume that the Gaussian process is defined on a regular rectangular grid, i.e.,

$$\mathbf{D} = \{1, \ldots, l\}^d \tag{5.9}$$

The Euclidean distance is not really adapted on a regular grid, so we now switch to the Manhattan distance, i.e., $|x - y| = \sum_{l=1}^{d} |x_l - y_l|$ and a covariance function of the form:

$$R(h;\theta) = \exp\left\{-\theta\left(\sum_{l=1}^{d} |h_l|\right)^p\right\} \quad \text{with } 0 < p \leq 2 \tag{5.10}$$

with $h_l = x_l - y_l$, $l = 1, 2, \ldots, d$.

Since the major source of variability in CE comes from the lack of fit, i.e., the discrepancy between the responses predicted by the model (5.4) and the runs (the outputs of the numerical simulations or code), it is mandatory to choose designs that spread the points (at which we want to run the code) throughout the region of interest. Classical factorial designs may be extremely resistant to this mandate. Anyway, there are a number of these designs in literature, known as *space-filling designs*, that meet the desired requirements. A design with this feature is obtained by simple random sampling or by stratified random sampling (if a portion of the experimental region needs to be explored with more refined investigations). One very popular design among CE users can be generated by applying the latin hypercube (LH) sampling proposed by McKay et al. (1979) in order to lower the variance of the unbiased (or asymptotically unbiased) estimator of $E[Y]$; i.e., the goal is to achieve the estimator with the smallest variance. However, while LH sampling has nice marginal properties (the points are marginally spread uniformly over the values of each input variable), not all LH designs are properly space-filling.

The next section is devoted to comparing different LH designs from among the total set of designs by considering a fixed number of variable levels.

5.4 Comparing Different Designs: Cases $2 \times 2, 3 \times 3, 4 \times 4$

In this section, we focus on finding classes of latin hypercube (LH) designs that have the same prediction features. We want to convince the reader that the choice of a LH design should made very carefully, and that such a choice must not be simply based on generic combinatorial or geometric arguments that are not directly related to the actual statistical computation of interest.

For the sake of the present discussion, we deliberately ignore the effects of estimating unknown model parameters. We focus on studying the variance of the Gaussian linear prediction as a function of the design, given that all parameters are known.

According to the Bayesian methodology for designing and analyzing CEs, the comparison is carried out on the variance of the conditional expectation (for a detailed and exhaustive discussion, see Santner et al. 2003). In Sack et al. 1989a, 1989b, the prediction of the unknown output $Y(\mathbf{x})$ occurs at a new site $\mathbf{x}_0$, i.e., at a generic point. In this chapter, we prefer to consider the prediction of the grand total $Y_+ = \sum_{i=1}^{n} Y(\mathbf{x}_i)$ of process variables for the training data, just as geostatisticians do (Krige 1951; Cressie 1993; Cressie 1997 and Matheron 1971).

According to the assumption of a conjoint multivariate normal distribution, see Sect. 5.2, the expected value and the variance of the grand total may be computed in closed form. For the sake of simplicity, we consider that β is the null vector and that the process variance σ_Z^2 is unitary. This standardization does not affect our conclusion. Under these circumstances, the total is a normal distribution with zero mean and a covariance matrix of $\{\sigma_{i,j}\}_{i,j=1,2,\dots,l^d}$, where $\sigma_{i,j}$ is the covariance between any two sets of training data $Y(\mathbf{x}_i)$ and $Y(\mathbf{x}_j)$. Therefore, $Y_+ \sim N(0, \sigma_+^2)$, where σ_+^2 is the variance of the total, comprising l^{2d} covariances (this number of covariances includes l^d variances).

Given the rectangular grid (5.9), $l!$ different LH designs are available. Let $\{Y_i^{(\mathrm{LH})_j}\}$ be the set of l training points in the LH_j design, $j = 1, 2, \dots, l!$. The distribution of the normal vector $(Y_+, Y_1^{(\mathrm{LH})_j}, Y_2^{(\mathrm{LH})_j}, \dots, Y_l^{(\mathrm{LH})_j})$ has zero mean and a covariance matrix of $\Sigma = \begin{pmatrix} \sigma_+^2 & \sigma_{+,(\mathrm{LH})_j} \\ \sigma'_{+,(\mathrm{LH})_j} & \Sigma_{(\mathrm{LH})_j} \end{pmatrix}$, where $\sigma_{+,(\mathrm{LH})_j}$ is the row vector of the covariances between the total and the point in the LH design considered, and $\Sigma_{(\mathrm{LH})_j}$ is the covariance matrix of the cited design.

The predictor $\hat{Y}_+^{(\mathrm{LH})_j} = \mathrm{E}(Y_+ | Y_1^{(\mathrm{LH})_j}, Y_2^{(\mathrm{LH})_j}, \dots, Y_l^{(\mathrm{LH})_j})$ is BLUP, and its mean square prediction error: (MSPE):

$$\mathrm{MSPE}\left[\hat{Y}_+^{(\mathrm{LH})_j}\right] = \mathrm{E}\left[\left(\hat{Y}_+^{(\mathrm{LH})_j} - Y_+\right)^2\right] \tag{5.11}$$

is the statistical index chosen in this chapter in order to compare different LH designs with the same number of training points. Given the orthogonality between $\hat{Y}_+^{(LH)_j} - Y_+$ and $\hat{Y}_+^{(LH)_j}$, the MSPE of (5.11) may be written as:

$$\text{MSPE}\left[\hat{Y}_+^{(LH)_j}\right] = \text{E}\left[(Y_+)^2\right] - \text{E}\left[\left(\hat{Y}_+^{(LH)_j}\right)^2\right]. \qquad (5.12)$$

Therefore, if the main reason that leads to the particular choice of an LH design among the $l!$ ones with the same number of points is to minimize the MSPE, we compare the different MSPEs of the designs in order to find the one that yields the predictor with maximum variance. In other words, the displacement of the LH design points must provide the predictor $\hat{Y}_+^{(LH)_j}$ that has maximum variance because of the relationship caused by the covariance (remember that since both $\hat{Y}_+^{(LH)_j}$ and Y_+ have means of zero, their variances overlap the second-order moment, and so the right side of (5.12) is the difference between their variances).

Finally, the best LH design—according the rule of maximum predictor variance—is the one that produces:

$$\max_{1 \leq j \leq l!} \text{var}\left[\hat{Y}_+^{(LH)_j}\right] = \max_{1 \leq j \leq l!} \sigma_{+,(LH)_j} \Sigma_{(LH)_j} \sigma'_{+,(LH)_j}. \qquad (5.13)$$

We compared all of the LH designs with two variables and $l = 2$, 3 and 4 levels. The predictor variances (5.13) were computed with the software CoCoA (Computations in Commutative Algebra), a freely available system for symbolic computation with multivariate polynomials. In fact, the closed form expression of the predictor variance is a rational function in the covariances with rational coefficients. As such, it is suitable for symbolic exact computation. Computation with the exponential model for the covariance was done using the software R.

First, we consider the trivial case with two factors and two levels of each factor. There are only two LH designs, and they coincide with the two regular fraction of a 2^2 factorial design. Because of the symmetry imposed by the covariance structure, the two designs have the same prediction variance.

5.4.1 3 × 3 LH Designs

When there are three levels, the situation is a bit more interesting. There are six LH designs, and two clusters are imposed by the symmetry, see Fig. 5.2. The former contain the two diagonal ones (1 and 4), and the latter contain all the remaining LH designs (2, 3, 5, 6).

The two LH designs with diagonal arrays of design points are not usually considered true space-filling designs; these designs are in the second cluster instead. This classification is not just qualitative—it is confirmed by an actual comparison between the respective predictor variances.

Fig. 5.2 Graphical representation of the array of three
design points in each of the
six LH designs with two variables and three levels

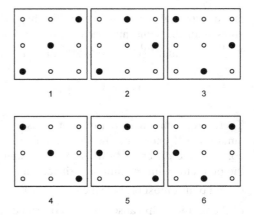

The structure of the covariance matrix Σ among the nine original points is:

$$\Sigma = \begin{pmatrix} 1 & \rho(1) & \rho(2) & \rho(1) & \rho(2) & \rho(3) & \rho(2) & \rho(3) & \rho(4) \\ \rho(1) & 1 & \rho(1) & \rho(2) & \rho(1) & \rho(2) & \rho(3) & \rho(2) & \rho(3) \\ \rho(2) & \rho(1) & 1 & \rho(3) & \rho(2) & \rho(1) & \rho(4) & \rho(3) & \rho(2) \\ \rho(1) & \rho(2) & \rho(3) & 1 & \rho(1) & \rho(2) & \rho(1) & \rho(2) & \rho(3) \\ \rho(2) & \rho(1) & \rho(2) & \rho(1) & 1 & \rho(1) & \rho(2) & \rho(1) & \rho(2) \\ \rho(3) & \rho(2) & \rho(1) & \rho(2) & \rho(1) & 1 & \rho(3) & \rho(2) & \rho(1) \\ \rho(2) & \rho(3) & \rho(4) & \rho(1) & \rho(2) & \rho(3) & 1 & \rho(1) & \rho(2) \\ \rho(3) & \rho(2) & \rho(3) & \rho(2) & \rho(1) & \rho(2) & \rho(1) & 1 & \rho(1) \\ \rho(4) & \rho(3) & \rho(2) & \rho(3) & \rho(2) & \rho(1) & \rho(2) & \rho(1) & 1 \end{pmatrix} \quad (5.14)$$

where $\rho(h)$, $h \in \{1,2,3,4\}$, is the correlation coefficient, which depends only on the
distance $|h|$ between any pair of points $\mathbf{x}_i$ and $\mathbf{x}_j$ (the process variance σ_Z^2 is unitary
according to the assumptions).

The variance of the grand total is:

$$\sigma_+^2 = 16 + 24\rho(1) + 28\rho(2) + 16\rho(3) + 4\rho(4). \quad (5.15)$$

The covariance matrices $\Sigma_{(LH)_{1-4}}$ for LH numbers 1–4 and $\Sigma_{(LH)_{2-3-5-6}}$ are:

$$\Sigma_{(LH)_{1-4}} = \begin{pmatrix} 1 & \rho(2) & \rho(4) \\ \rho(2) & 1 & \rho(2) \\ \rho(4) & \rho(2) & 1 \end{pmatrix}$$

$$\Sigma_{(LH)_{2-3-5-6}} = \begin{pmatrix} 1 & \rho(3) & \rho(3) \\ \rho(3) & 1 & \rho(2) \\ \rho(3) & \rho(2) & 1 \end{pmatrix} \quad (5.16)$$

Figure 5.3 shows a plot of the relative difference between the variances of the grand
total predictors based on the two cluster designs as a function of the parameter θ.
We use this relative difference as an efficiency index. The LHs numbered 2, 3, 5

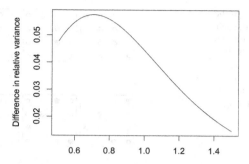

Fig. 5.3 Comparisons of grand total predictors based on two different LH designs: the data refers to the relative difference between the variances of the two grand total predictors as a function of the covariance function parameter θ with smoothing parameter $p = 1$

and 6 (see Fig. 5.2) lead to a slightly more efficient predictor of the grand total (i.e., a smaller MSPE), but the efficiency tends to vanish as the parameter θ increases (as θ increases, correlation decreases).

5.4.2 4 × 4 LH Designs

When four levels are considered for each factor, there are $4! = 24$ LH designs available, all of which are depicted in Fig. 5.4. The computations were only performed for eight designs, each one representing a unique cluster (the other designs can be obtained by circular permutation of the design points, and they give the same prediction variance for the total).

In Fig. 5.5 there are the plot of the relative variances of the predicted value of the grand total according to the pinpointed LH designs: LHs 1, 2, 3, 4 in top display and 8, 10, 11, together with the design consisting of the four corners in the bottom display. Also in such a case the different behaviours of the predictors based on the LH designs are not noteworthy. A major discrepancy may be detected in the comparison with the predictor based on a factorial design with the same number of the design points as well. A possibility would be a $4^{(2-1)}$ design, such as the four corners design we have used. The comparison is in favour of any LH designs, even considering the not truly space filling ones. This picture may be an assist in choosing the proper LH design among the many available according to the criterion of maximum variance of the grand total predictor.

These comparisons were performed in order to aid the selection of the design points. In fact, it can be rather tricky to define the appropriate design according to the universal Kriging model (5.4): if the choice of the design is made to fulfil the goodness requirements of the regression part of the model (i.e., $\beta'\mathbf{f}(\mathbf{x})$), it is advisable to use a classical factorial design; on the other hand, if the design should favor the stochastic process component $Z(\mathbf{x})$, a space-filling design (like the LH designs) is better.

Fig. 5.4 Representation of the
array of design points in each
of the 4! LH designs with two
variables and four levels. It
is evident that there are are
only eight relevant designs,
because the remaining designs
can be obtained by rotating
these eight designs

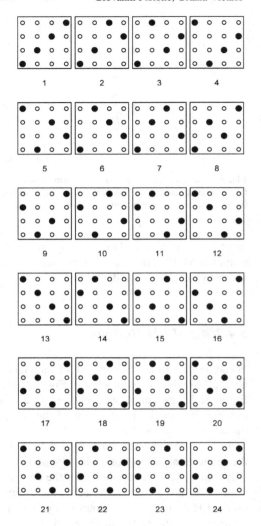

5.5 Conclusions

We have discussed the Kriging model, where the covariance depends on the Manhattan distance between the points on a regular grid of location. After analyzing the 3×3 and 4×4 cases, we can draw the following conclusions:

1. The efficiency of an LH design in terms of the relative mean square error of the conditional expectation step is better than those of other designs with the same number of points, even those that are optimal for regression problems.

Fig. 5.5 Relative variance of the predicted value of the grand total for the LHs 1, 2, 3, 4 (*top*), 8, 10, 11 and the design consisting of the four corners (*bottom*). The line types, in order, are: *solid, dashed, dotted, dotdash*. The parameter θ ranges from 0.5 to 2.5 and $p = 1.2$

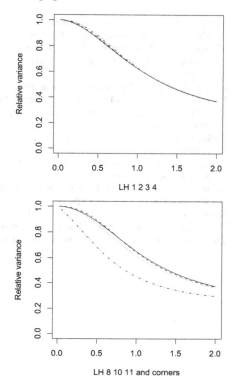

2. All LH designs have essentially the same efficiency for small grids (3×3 and 4×4 here). This contrasts with the current suggestion: to look for an LH with good spatial coverage.
3. We suggest that a preliminary analysis of this type should be performed in order to single out a suitable class of LH design and randomize the LH in this class, rather then to choose a good-looking design a priori.
4. Up to the dimensions considered, symbolic exact computations are feasible and could provide some extra insight.

References

Box, G.E.P., Draper, N.R.: Empirical Model-Building and Response Surface. Wiley, New York (1987)

Cressie, N.A.: Statistics for Spatial Data. Wiley, New York (1993)

Cressie, N.A.: Spatial prediction and ordinary kriging. Math. Geol. **20**(4), 407–421 (1997)

Kleijnen, J.: Statistical Tools for Simulation Practitioners. Marcel Dekker, New York (1987)

Krige, D.G.: A statistical approach to some mine valuations and allied problems at the Witwatersrand. Master's Thesis, University of Witwatersrand, Johannesburg (1951)

Matheron, G.: The theory of regionalized variables and its applications. Les Cahiers du Centre de Morphologie Mathematiques de Fontainbleau (France) 5 (1971)

Mc Kay, M.D., Beckman, R.J., Conover, W.J.: A comparison of three methods for selecting values of input variables in the analysis of out put from a computer code. Technometrics **21**, 239–245 (1979)

Loh, W.-L.: On latin hypercube sampling. Ann. Stat. **28**, 2058–2080 (1996)

Park, S., Fowler, J.W., Mackulak, G.T., Keats, J.B., Carlyle, W.M.M.: D-optimal sequential experiments for generating a simulation-based cycle time-throughput curve. Oper. Res. **50**, 6, 981–990 (2002)

Sacks, J., Schiller, S.B., Welch, W.J.: Design for computer experiments. Technometrics **31**, 41–47 (1989a)

Sacks, J., Welch, W.J., Mitchell, T.J., Wynn, H.P.: Design and analysis of computer experiments. Stat. Sci. **4**, 409–423 (1989b)

Santner, T.J., Williams, B.J., Notz, W.I.: The design and analysis of computer experiments. Springer-Verlag, New York (2003)

Sasena, M.J., Papalambros, P., Goovaerts, P.: Exploration of metamodeling sampling criteria for constrained global optimization. Eng. Optimiz. **34**, 3, 263–278 (2002)

Zienkiewicz, O.C.: The Finite Element Method in Engineering Science. McGraw-Hill, New York (1971)

Chapter 6
New Sampling Procedures in Coordinate Metrology Based on Kriging-Based Adaptive Designs

Paola Pedone, Daniele Romano, and Grazia Vicario

Abstract This chapter describes an interesting case of process innovation generated by transferring two statistical technologies from their native application fields to a different one. The technologies are prediction by *Kriging models* and *sequential experiments*, originally developed for geostatistics applications and clinical trials, respectively. The combination of the two, i.e., sequential experiments driven by Kriging predictions, has been successfully applied in coordinate metrology. The latter is a vast technical sector, widespread in industry, devoted to assessing product compliance to geometrical specifications by measuring a set of point coordinates on the part to be inspected. Preliminary results indicate that this technology transfer has produced a remarkable improvement in the performance of the measurement process, in terms of both quality and productivity.

6.1 Introduction

An essential problem for researchers and engineers is the reconstruction of a surface of a manufactured product on the basis of a set of measured points. The set of points is obtained by a measurement process performed by a coordinate measuring

Paola Pedone
INRIM (Italian National Research Institute of Metrology)
Strada delle Cacce 91, 10135 Torino, Italy
e-mail: p.pedone@inrim.it

Daniele Romano
Department of Mechanical Engineering, University of Cagliari
Piazza d'Armi, 1, 09123 Cagliari, Italy
e-mail: romano@dimeca.unica.it

Grazia Vicario
Politecnico di Torino, Department of Mathematics
Corso Duca degli Abruzzi 24, 10129 Torino, Italy
e-mail: grazia.vicario@polito.it

Pasquale Erto, *Statistics for Innovation*
ISBN 978-88-470-0814-4, © Springer 2009

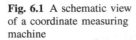

Fig. 6.1 A schematic view
of a coordinate measuring
machine

machine (CMM). CMMs are finding increasing use, both online for production control and offline for the inspection of finished products. They can work under either operator control (for one-at-a-time jobs) or computer control (for repetitive tasks).

A CMM inspection (see Fig. 6.1) is typically performed by sequentially logging the coordinates of points where a ball-end touch-fire probe contacts the surface of the piece under consideration and sends back a set $S = \{x_1, x_2, \ldots, x_n\}$ of Cartesian coordinates (in one or more dimensions: $x_j \in \mathbb{R}^d$, with $1 \leq d \leq 3$ and $j = 1, 2, \ldots, n$) pertaining to contact points between the device (touch probe) and the surface explored; polar and cylindrical coordinates are also used whenever convenient.

Point-by-point exploration is a sampling process, and the size of the sample is necessarily limited by time and cost constraints. Moreover, the coordinates returned by the machine are affected by random errors whose effects on the final result of the control have to be assessed. These facts make the evaluation of form errors with CMMs quite a challenging statistical problem, especially when complex controls are involved. To make the inferential problem even more difficult, form errors, as defined by tolerancing standards (ASME Y14.5.1M, 1994), depend heavily on extreme values of the deviations in form over the surface, so a full-field inspection is virtually required. As an example, straightness is defined as "the minimum distance between two parallel lines enclosing the actual feature" (minimum zone criterion). Hence, a few points, the most "outward" and "inward" points, determine straightness, whereas the others are irrelevant; see Fig. 6.2. In statistical terms, this is equivalent to making inference on parameters that are dependent on tail units of the population of form deviations, which are unlikely to occur in a small sample. In spite of this, industrial practice envisages the use of very simple sampling methods (uniform, stratified, random) and the evaluation of form error by deterministic methods based only on the sample data, thus neglecting the statistical nature of the estimation problem.

Fig. 6.2 Form error depends
only on a few extreme points
in the surface pattern

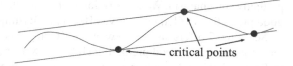

critical points

Fig. 6.2 Form error depends only on a few extreme points in the surface pattern

Figure 6.3 shows an example of the two most commonly used methods. The orthogonal least squares (OLS) method encloses the points between a pair of lines, both of which are parallel to the orthogonal least squares line (minimizing the sum of the squared orthogonal deviations between the probed points and the line). Straightness is measured as the minimum distance between two such lines. The other method, the convex hull method (CH), assesses the minimal convex set containing the probed points. For each side of the convex hull, orthogonal distances between that side and each vertex are computed and the largest is saved. Straightness is then obtained as the smallest of these largest distances.

The purely deterministic evaluation of form errors provided by both methods is naturally prone to underestimating the actual error (Dowling et al. 1997): points of the actual surface external to the region enclosed by the parallel lines would generally produce a larger error. In several instances, this downward bias can be high enough to accept parts which should have been rejected.

One way to overcome these inherent difficulties is to exploit some additional information on the actual surface to be inspected: in-process information and a priori information. The differences depend on whether the additional information is collected during the inspection stage or before (e.g., knowledge of the manufacturing processes applied, preliminary detailed inspection of a few parts). In this chapter, we consider in-process information, which is only poorly explored in the technical literature (Badar et al. 2003; Edgeworth and Wilhelm 1999). In particular, we focus on the construction of a sequential inspection plan where information collected up the current probed point is exploited to select the next point. The plan is then terminated on reaching a stopping criterion.

Since the construction of the sequential design relies heavily on the prediction of the surface pattern at each step of the procedure, a good design requires a good

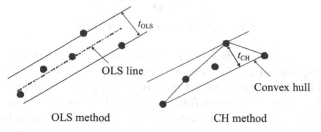

t_{OLS}

OLS line

OLS method

t_{CH}

Convex hull

CH method

Fig. 6.3 The main sample-based methods used for form error computation (five points have been probed on a nominally straight line). *Left*: orthogonal least squares (OLS). *Right*: convex hull (CH)

predictive mechanism. We use Kriging models for their recognized ability to provide good predictions (Kleijnen and van Beers 2004; Simpson et al. 2001).

Section 6.2 offers an overview of the Kriging model and on the issue of parameter estimation. The good predicting ability of Kriging models is then demonstrated in Sect. 6.3 through a comparison with regression models in a selected test case. Section 6.4 is devoted to a description of the adaptive approach for generating sequential sampling plans. Its application is then illustrated by two case studies based on real CMM measurements. A final discussion encouraging the use of the proposed approach concludes the chapter.

6.2 Kriging Models

Kriging models are named after a South African engineer, Daniel G. Krige (1951), who first referred to them when analyzing mining data. Facing the problem of having to make accurate predictions of response based on a rather small set of spatial data, he put the reasonable idea that response values that are spatially close together are much more alike than values that are more distant (smoothness of the response) in statistical terms. Consequently, when predicting at an untried location, observations that are closer to it should exert more influence on the prediction. His intuition was further developed in geostatistics by other authors (Matheron 1971; Cressie 1993 and 1997; Goovaerts 1997). Then, at the end of the 1980s, a group of statisticians (Sachs et al. 1989a,b) set up a framework that used Kriging to model the deterministic output of a computer program.

Let $D \subset \mathbb{R}^d$ be the region where we want to predict the response y observed at a set of experimental points $\mathbf{x}^n = (\mathbf{x}_1, \mathbf{x}_2, \ldots, \mathbf{x}_n)$, with $\mathbf{x}_i \in D$ for $i = 1, 2, \ldots, n$. Response $y(\mathbf{x})$, for $\mathbf{x} \in D$, is considered a realization of a Gaussian random process $Y(\mathbf{x})$:

$$Y(\mathbf{x}) = \beta' \mathbf{f}(\mathbf{x}) + Z(\mathbf{x}) \tag{6.1}$$

where $\mathbf{f}(\mathbf{x}) = (f_1(\mathbf{x}), f_2(\mathbf{x}), \ldots, f_m(\mathbf{x}))'$ is a set of specified trend functions and $\beta = (\beta_1, \beta_2, \ldots, \beta_m)'$ is a set of coefficients. $Z(\mathbf{x})$ is a Gaussian random process with zero mean and stationary covariance over D, so that:

$$E[Y(\mathbf{x})] = \beta' \mathbf{f}(\mathbf{x})$$
$$\text{Cov}(Y(\mathbf{x}), Y(\mathbf{x}+\mathbf{h})) = \sigma_Z^2 R(\mathbf{h}; \theta) \tag{6.2}$$

where σ_Z^2 is the process variance, R is the stationary correlation function (SCF), which only depends on the displacement vector $\mathbf{h}$ between any pair of points in D and on a set of parameters θ. Model (6.1) is known as universal Kriging. A convenient choice for the correlation function is found within the power exponential family (the most common model adopted in the computer experiment literature):

$$R(\mathbf{h}; \theta) = \prod_{l=1}^{d} \exp\left\{-\theta_l |h_l|^p\right\} = \exp\left\{-\sum_{l=1}^{d} \theta_l |h_l|^p\right\} \quad \text{with } 0 < p \le 2 \tag{6.3}$$

where $\theta = (\theta_1, \theta_2, \ldots, \theta_d, p)'$, θ_l are positive scale parameters and p is a common smoothing parameter. Parameter θ_l describes how rapidly the correlation decays in direction l with increasing distance h_l. The assumption in (6.3), that the positive correlation between outputs diminishes with increasing distance between the input sites, is a formalization of Krige's original idea. Notice that if $\theta_l = \theta \ \forall l$, the correlation depends only on the distance $|\mathbf{h}|$ between any pair of points $\mathbf{x}$ and $\mathbf{x} + \mathbf{h}$ (isotropic SCF). Parameter p describes the pattern of correlation decay. When $p = 2$ we have the (inappropriately) termed "Gaussian SCF," which is suitable for very smooth, infinitely differentiable, responses. In general, the less smooth the response, the lower p is. The practitioners in computer experiments restrict the choice of the SCF to functions with the desired smoothness properties; for a formal treatment of the subject, see Abrahamsen (1997).

Kriging modeling lends itself to a sound theoretical framework for predicting the output. Let $\mathbf{Y}^n = (Y(\mathbf{x}_1), Y(\mathbf{x}_2), \ldots, Y(\mathbf{x}_n))'$ be the set of process variables at x^n. Then, under the hypothesis (consistent with (6.1) and (6.2)) that the joint random variable $(Y(\mathbf{x}_0), Y(\mathbf{x}_1), Y(\mathbf{x}_2), \ldots, Y(\mathbf{x}_n))$ is a multivariate normal, the mean of $Y(\mathbf{x})$ at the untried point $\mathbf{x}_0$ conditional on the process data, $\hat{Y}_0 = \mathrm{E}(Y(\mathbf{x}_0)|\mathbf{Y}^n)$, is:

$$\hat{Y}_0 = \mathbf{f}_0'\boldsymbol{\beta} + \mathbf{r}_0'\mathbf{R}^{-1}(\mathbf{Y}^n - \mathbf{F}\boldsymbol{\beta}) \tag{6.4}$$

where $\mathbf{f}_0$ is the $m \times 1$ vector of the trend functions in $\mathbf{x}_0$; $\mathbf{F}$ is the $n \times m$ matrix $\{f_j(\mathbf{x}_i)\}_{\substack{i=1,\ldots,n \\ j=1,\ldots,m}}$ of the trend functions computed in $(\mathbf{x}_1, \mathbf{x}_2, \ldots, \mathbf{x}_n)$, $\mathbf{r}_0$ is the correlation vector $(R(\mathbf{x}_0 - \mathbf{x}_1), \ldots, R(\mathbf{x}_0 - \mathbf{x}_n))'$, and $\mathbf{R}$ is the $n \times n$ correlation matrix whose (i, j) element is $R(\mathbf{h}_{ij} = \mathbf{x}_i - \mathbf{x}_j)$.

The predictor $\hat{Y}_0 = \mathrm{E}(Y(x_0)|\mathbf{Y}^n)$ is the best linear unbiased predictor (BLUP) of $Y(\mathbf{x}_0)$, because it minimizes the mean squared prediction error $\mathrm{E}[(\hat{Y}_0 - Y_0)^2]$, and it is also the only one. Predictor (6.4) can also be regarded as the weighted linear combination of the observations that minimizes the mean squared prediction error under the constraint of being unbiased, $\mathrm{E}[\hat{Y}_0] = \mathrm{E}[Y_0]$.

The predictor is the sum of the regression term and the correction term $\mathbf{r}_0'\mathbf{R}^{-1}(\mathbf{Y}^n - \mathbf{F}\boldsymbol{\beta})$. The correction term can be interpreted as being a linear combination of the residuals of the fitted regression model that forces the predictor to interpolate the observed data. It is the key element of Kriging, as the correlation structure acts through it. It generally produces good predictions even when the regressive term is unable to capture the actual trend. In fact, the model typically used in geostatistics, referred to as ordinary Kriging, reduces the regressive term to a constant (i.e., $\boldsymbol{\beta}'\mathbf{f}(\mathbf{x}) = \beta$) without suffering losses in prediction fidelity (Sacks 1989). Therefore, we use ordinary Kriging both because we assume that no knowledge on the surface error is available a priori to drive the choice of β, and because this allows a faster inspection plan, as a lower number of parameters will be estimated at each step.

The mean squared prediction error:

$$\mathrm{MSPE}\left[\hat{Y}_0\right] = \mathrm{E}\left((\hat{Y}_0 - Y(\mathbf{x}_0))^2\right) = \sigma_z^2\left(1 - \mathbf{r}_0'\mathbf{R}^{-1}\mathbf{r}_0\right), \tag{6.5}$$

usually called the Kriging variance, is a measure of prediction uncertainty. It is large when $\mathbf{x}_0$ is far from the experimental points and small when it is close to them. Due to the interpolatory nature of Kriging, it vanishes at the experimental points.

However, Eqs. 6.4 and 6.5 only hold when β and $R(\mathbf{h}; \theta)$ are known, which is hardly ever the case in practical situations. When β is unknown, its generalized least squares estimator $\hat{\beta} = (\mathbf{F}'\mathbf{R}^{-1}\mathbf{F})^{-1}\mathbf{F}'\mathbf{R}^{-1}\mathbf{Y}^n$ must replace β in (6.4) to yield the new predictor. In this case, the Kriging variance is larger than (6.5) and becomes:

$$\mathrm{E}\left(\left(\hat{Y}_0 - Y(\mathbf{x}_0) \right)^2 \right) = \sigma_z^2 \left(1 - \mathbf{r}_0'\mathbf{R}^{-1}\mathbf{r}_0 + \mathbf{c}_0' \left(\mathbf{F}'\mathbf{R}^{-1}\mathbf{F} \right)^{-1} \mathbf{c}_0 \right) \qquad (6.6)$$

with $\mathbf{c}_0 = \mathbf{f}_0 - \mathbf{F}'\mathbf{R}^{-1}\mathbf{r}_0$.

The most common case is when the vector θ in $R(\mathbf{h}; \theta)$ is unknown. It can then be estimated by maximum likelihood, cross-validation, or posterior mode (for a thorough reading see Santner et al. 2003). The maximum likelihood estimate is:

$$\hat{\theta}_{ML} = \arg(\min(n \log \hat{\sigma}_z^2(\theta) + \log(\det(\mathbf{R}(\theta))))) \qquad (6.7)$$

where:

$$\hat{\sigma}_Z^2(\theta) = \left(\mathbf{y}^n - \mathbf{F}\hat{\beta}(\theta) \right)' \mathbf{R}(\theta)^{-1} \left(\mathbf{y}^n - \mathbf{F}\hat{\beta}(\theta) \right) \qquad (6.8)$$

is the MLE of the process variance. In (6.8), $\mathbf{y}^n$ is the vector of the observations, and the estimate $\hat{\beta}(\theta)$ is provided by generalized least squares. The predictor obtained by plugging the estimates $\hat{\mathbf{r}}_0 = \mathbf{r}_0(\hat{\theta}_{ML})$ and $\hat{\mathbf{R}} = \mathbf{R}(\hat{\theta}_{ML})$ into (6.4) is termed the *empirical best linear unbiased predictor* (EBLUP). However, the predictions are no longer linear in the observations, as $\hat{\mathbf{r}}_0$ and $\hat{\mathbf{R}}$ can have a highly nonlinear dependence on observations. Another notable consequence of using the EBLUP is that (6.6) underestimates the prediction variance as it does not account for the extra variability transmitted to $\hat{\mathbf{r}}_0$, $\hat{\mathbf{R}}$ and $\hat{\beta}$ by $\hat{\theta}$. Possible ways of overcoming this problem include resorting to an empirical estimate of the variance. Den Hertog et al. (2006) use parametric bootstrap, and Kleijnen and van Beers (2004) use cross-validation and jackknife.

6.3 Prediction Capability: Kriging vs. Regression

In this section, we compare the prediction capabilities of Kriging models with those of standard regression models using a test case selected ad hoc. Regression is the most common *global* model for statistical interpolation. In contrast, Kriging models, though formally global (see Eq. 6.1), can be regarded as *local* due to the flexible spatial correlation mechanism that they are based on. Different behavior can therefore be expected. The test case is peculiar in that the response exhibits fast and complex variations in one input variable, but very slow variations in the remaining two variables. The test case is selected according to a wider research activity aimed at designing an optical profilometer based mainly on a stochastic simulation

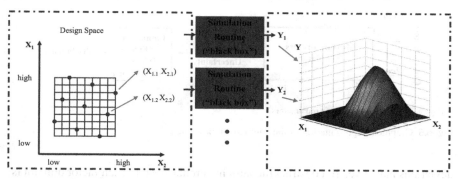

Fig. 6.4 Graphical scheme of the simulation experiment. The third factor is omitted for ease of representation

model of the whole measurement process (see Fig. 6.4). The design, carried out at the Dept. of Mechanical Engineering of the University of Cagliari, was performed using a new robust design procedure (Giovagnoli and Romano 2008), and culminated with an innovative prototype that exhibits improved performance and reduced production costs (Pedone 2006; Baldi et al. 2006). The simulator was validated by physical trials (Baldi et al. 2006; Baldi and Pedone 2005).

The test case involves experiments with three control factors. The first factor is the size of the mirror displacement (p); the other two are the scatter (standard deviation) in the mirror displacement (σ_x) and the scatter in the gray-tone levels of the pixels of digital images acquired by a video camera (σ_{CCD}). The experimental region is the hypercube $(0.1\,\mu m, 0.3\,\mu m) \times (0.005\,\mu m, 0.015\,\mu m) \times (0.008\,\mu m, 0.016\,\mu m)$. The comparison is performed using two different sampling strategies: full factorial designs and latin hypercube (LH) designs.

The response of interest is the measurement uncertainty, computed as the sample standard deviation for 10^4 replications performed for each design treatment (i.e., combination of levels of the controlled factors). In each replication, all of the uncontrolled errors associated with the mirror displacement and the gray-tones of the image pixels are simulated by drawing from their assumed distributions (zero-mean normal with standard deviations σ_x and σ_{CCD}, respectively). After the experiments have been conducted, response models are estimated using both regression and Kriging. Models are used to obtain predictions at untried points within the experimental region. Predictions are then assessed by comparing them with the corresponding values obtained by the simulator. At the prediction stage, the response is evaluated using 10^6 replications of the stochastic simulator, i.e., with more precision than for the training data. The whole test procedure is represented in Fig. 6.5.

Three different sample sizes are considered for the training designs, namely $n = 27$, 125, and 225. The full factorial experiments involve 3^3, 5^3 and 9×5^2 designs, respectively. In the first two experiments, all factors are given three and five equispaced levels covering the factor ranges. In the last experiment, the mirror displacement—the most influential factor—has nine levels, while the two standard

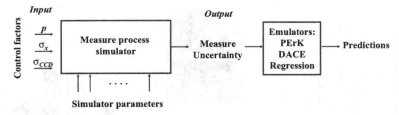

Fig. 6.5 Graphical representation of the comparison procedure

deviations have five. To construct the latin hypercube design, each factor domain is divided into n strata with equal marginal probabilities $1/n$, and one level is selected from each stratum by random sampling. Then the n design points are obtained by random sampling without replacement from the set of levels of each factor. The distinctive features of LH designs are the wide variety of factor levels and the freedom to choose any size of experiment.

Regression models are polynomials; the degree of the polynomial is constrained by the type of experimental design used. They are estimated by the stepwise procedures available in the *Minitab* software. For Kriging models, both universal (model (6.1)) and ordinary Kriging (i.e., $Y(\mathbf{x}) = \beta + Z(\mathbf{x})$) are considered. The models are estimated using two software packages available on the web: PErK (*Parametric Empirical* Kriging) and DACE (*Design and Analysis of Computer Experiments*). The latter is also compatible with the MATLAB code. The two packages implement different correlation functions: the full exponential family and the Matèrn I[1] for PErK, and only the Gaussian (exponential family with the restriction $p = 2$) for DACE.

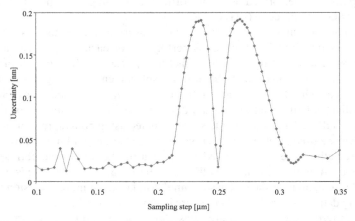

Fig. 6.6 Uncertainty vs. mirror displacement: the function is computed for several input levels with the factors σ_x and σ_{CCD} held constant

---

[1] The Matèrn I correlation function is: $R(h) = \prod_{j=1}^{d} \frac{1}{\Gamma(v)2^{v-1}} \left(\frac{2\sqrt{v}|h_j|}{\theta_j} \right)^{v_j} K_v \left(\frac{2\sqrt{v}|h_j|}{\theta_j} \right)$ where θ_j is the scale parameter and K_v is the modified Bessel function of order $v > 0$.

The parameters of the correlation functions are estimated by an MLE procedure in both the packages.

The particular relationship between the mirror displacement and the response, which presents marked variations with local maxima and minima, is shown in Fig. 6.6. Factors σ_x and σ_{CCD}, which barely affect the response, are held at fixed values in the diagram. To smooth the response pattern, models are estimated for the natural logarithm of the response.

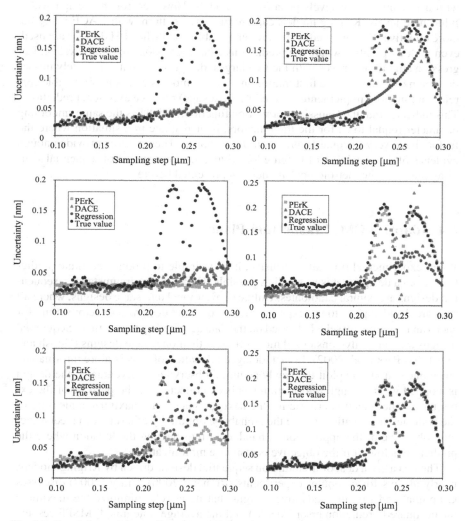

Fig. 6.7 Comparison of the prediction capabilities of regression and Kriging models for six combinations of design type and size used for model estimation. *Left column*: factorial designs; *right column*: LH designs. *Top row*: 27 points; *middle row*: 125 points; *bottom row*: 225 points

The test sample used to assess the predictive capabilities of the models comprises 201 points. The levels of mirror displacement are uniformly partitioned across the factor range, while the levels of σ_x and σ_{CCD} are selected at random within their ranges. Figure 6.7 displays a comparison between the actual and predicted responses for the six combinations of design type (factorials, LH) and size $(27, 125, 225)$. The graphs are two-dimensional, with only the dominant factor, p, explicitly represented on the x-axis. In fact, the effects of σ_x and σ_{CCD} are much smaller and introduce only tiny variations into the graphs compared to the pattern in Fig. 6.6. Factorial designs with a low number of levels penalise all models. However, for the design where p has nine levels, Kriging models capture the wavy pattern, with DACE being the more effective. Kriging predictions are remarkably good when LH designs are used, even with 27 points—where the predictions from regression are totally wrong. Regression slightly improves with the two largest design sizes, but completely misses the first maximum and the first minimum. Kriging models with a more flexible correlation function (implemented in the PErK package) produce excellent predictions. This indicates that the Gaussian correlation function, which has the same smoothing parameter (equal to 2) for the three factors, is inadequate in a situation where the factors have very unequal leverage on the response. These results provide enough evidence of the suitability of Kriging for driving the construction of sequential sampling plans for the metrological application presented before.

6.4 Adaptive CMM Inspection Plans

When experimental runs are expensive, it is desirable to produce accurate predictions from a design that is as small as possible. Under these circumstances, sequential designs are natural candidates. Unlike conventional one-stage designs, where all runs are decided prior to the experiment, in sequential designs the factor setting for each run is adaptively selected based on the data acquired up to that time. Sequential designs are generally considered more efficient than one-stage designs (Ghosh and Sen 1991; Park et al. 2002). In fact, delayed allocation of runs is more informative, since one can also exploit the newly acquired observations. Eventually, the design is stopped when enough information has been collected for the purpose of the experiment. The selection criteria for the next run may be to maximize a measure of the collected information, to use the alphabetic optimality criteria, or to meet a specific objective of the application at hand, e.g., by selecting the location where the prediction is lowest if the objective is response minimization.

There is a great deal of literature on sequential designs driven by Kriging models, nearly always for computer experiments (Romano 2006). Crary (2002) discusses G-optimal and I-optimal adaptive designs that iteratively minimize the maximum mean squared prediction error (MSPE) and the average (integrated) MSPE respectively. Jones et al. (1998) use a Bayesian approach to response optimization, where the next input site is chosen by a heuristic criterion maximizing the expected improvement in the search for an unconstrained optimum. The procedure is driven by the posterior density of the improvement function, conditional on all data available at each step. Variations of this method accommodating for the presence of constraints are proposed by Shonlau et al. (1998), while Williams et al. (2000) and Lehman et al. (2004) use a Bayesian sequential strategy that deals with noise variables to solve a robust design problem. Kleijnen and van Beers (2004) use a frequentist approach and select the next point based on where the prediction variance is at its maximum. For a comprehensive discussion of criteria for Kriging-based sequential designs, see Sasena et al. (2002).

These methods generally start with a space-filling design, like a latin hypercube sampling design, a distance-based design, or a uniform design (Santner et al. 2002; Fang et al. 2006); then, once the sequential procedure is activated, the estimates for the parameters of the correlation function of the Gaussian process are updated using some method (e.g., maximum likelihood, cross-validation, least squares) at each step or, to reduce the computational burden, after a few steps.

In this chapter, we use the sequential designs described in the flow diagram in Fig. 6.8; for a detailed description, see Pedone et al. (2008).

The procedure starts with a fixed-size design of n_0 equispaced points, including the extreme points of the domain to avoid extrapolation when predicting with Kriging models. In the case study, we take $n_0 = 4$, which is the minimum number of data

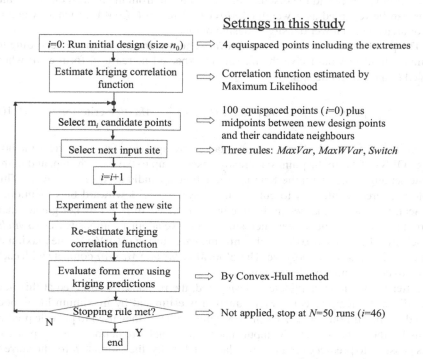

Fig. 6.8 Flow diagram of the general adaptive procedure for buiding sequential designs. The settings used in the two case studies discussed in this chapter are reported on the *right* of the diagram

points needed to estimate all of the parameters of the ordinary Kriging model. The design is then iteratively built up by adding one point at a time according to a pre-determined rule. The added design point is selected from a set of candidate points; the size of this set is allowed to vary from one step to another because the prediction reliability improves with design size (it is sensible to increase the number of candidate points with the step number). The augmentation rule, also used by Kleijnen and Van Beers (2004), is a dynamic one and allows us to focus on particular area of the domain if necessary.

During the sequential construction of the plans, two main criteria are taken into consideration. One is informative and one is problem-specific. The informative criterion is the maximum prediction variance (*MaxVar*), i.e., the next point x_i is selected on the basis of where the Kriging predictions are most uncertain:

$$x_i = x_{ij^*}, \quad j^* = \arg\left(\max_{1 \le j \le m_i} \text{Var}(\hat{y}_{ij}) \right) \quad MaxVar \text{ criterion} \qquad (6.9)$$

where $\hat{y}_{ij}$ is the Kriging prediction at the candidate x_{ij}, and m_i is the number of candidate points at step i. A variant of *MaxVar* is also considered by combining *MaxVar* with the ancillary criterion that wide areas of the domain should not be unexplored. According to this, the winning candidate is the one that maximizes the product of the prediction variance and the distance of the candidate from the nearest design point. This can be regarded as a weighted *MaxVar* criterion (*MaxWVar*) where weights favor areas with a lower density of design points.

The problem-specific criterion (*MaxInc*) chooses the next point x_i according to which candidate would give the maximum expected increase in form error when added to the current design:

$$x_i = x_{ij^*}, \quad j^* = \arg\left(\max_{1 \le j \le m_i} (\hat{t}_{ij} - t_i) \right) \quad MaxInc \text{ criterion} \qquad (6.10)$$

where t_i and $\hat{t}_{ij}$ are the form errors obtained by applying one numerical method (e.g., OLS or CH) to the point set already observed up to the $i-1$-th step, and to the same set augmented with the Kriging prediction at candidate x_{ij}, respectively. This criterion directly attempts to cut down the systematic downward bias mentioned in Sect. 6.1. Notice that when the maximum increase of $\hat{t}_{ij} - t_i$ is not greater than zero, the criterion is no longer interesting. Thus, we adopt a composite rule (*Switch*) which switches from *MaxInc* to the informative criterion (*MaxVar*) if the maximum expected increase is not positive. Therefore *MaxInc* and *MaxVar* compete with each other to choose the next design point.

After the winning candidate is designated, the response is observed at this new site. Then the Kriging correlation function is estimated by maximum likelihood, based on the current dataset. The new Kriging model is then used to provide predictions for the response over the input space. Since such predictions are inexpensive, it is possible to predict over a fine regular grid. Finally, the estimate $\hat{t}_i$ for the current form error is computed by applying OLS or CH methods to the large point sample comprising both the current experimental points and the freshly obtained predictions. In the case studies, we compute predictions for all current candidates and at all points in-between each candidate pair, and form errors are calculated by the CH method.

The final design size N is the step at which the stopping rule is met. However, in this study we do not apply any stopping rule; instead, we terminate the experiment at a size $N = 50$, i.e., after 46 cycles of the algorithm. In fact, the research is exploratory in nature, as it aims to assess the potential of new inspection plans for the particular application. Moreover, as the number of measured points per part is the main industrial constraint in CMM operations, a maximum budget of fifty points is a reasonable choice.

6.5 Application to Straightness and Roundness

In this section we describe the application of the sequential procedures to two case studies. The procedures are not applied online. In fact, the goodness of the approach is best tested if the true value of the form error is known. Thus, a preliminary large point sample is inspected over the surface, so that the computed form error, obtained by either the OLS or the CH method, is a good approximation of the true one. Then the sequential plan is iteratively created with the constraint that all points belong to the large sample. As the surface is inspected in a dense regular grid, the approximation made by replacing the next point, dictated by the relevant criterion, with the nearest grid point is indeed negligible. The accuracy of the estimated error at the i-th step is defined as:

$$A_i = \frac{\hat{t}_i}{t_{\text{true}}} \cong \frac{\hat{t}_i}{t_{\text{LS}}} \tag{6.11}$$

where t_{true} is the true unknown form error, t_{LS} is its large-sample approximation, and $\hat{t}_i$ is its estimate at the i-th step. In the following, the accuracy obtained from sequential procedures is compared with that obtained from simple fixed-size sampling methods. Among such methods, pure random sampling (PRS) and latin hypercube sampling (LHS) are considered here.

6.5.1 Case 1: Straightness

Point data are taken for a selected area on a rectangular plate. Measurements were made by scanning 62 parallel lines over the area with a step size of 0.5 mm (Buonadonna et al. 2007). The case study involves one line (line 34) over the rectangular area, comprising a total of 331 probed points. Large sample measurements are plotted in Fig. 6.9 as the lighter solid line. Note that only three points (R,S,T in Fig. 6.9) determine the straightness error ($t_{\text{LS}} = 53.4\,\mu\text{m}$). To be successful, the procedure should chase only those points. The application of the three previously defined rules leads to the sequential plans plotted in Fig. 6.9. Dots represent the design sites up

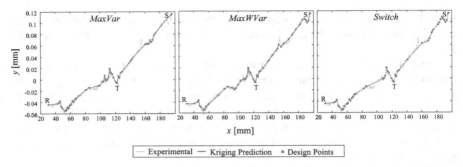

Fig. 6.9 Kriging-based sequential designs with $N = 50$ points for the straightness case study. Next-point selection rules are: *MaxVar* (*left*), *MaxWVar* (*middle*), *Switch* (*right*)

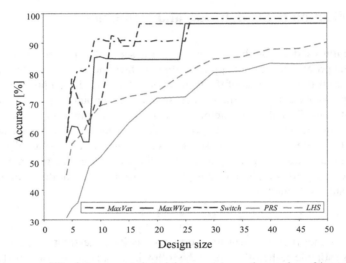

Fig. 6.10 Accuracy (%) of the computed straightness error vs. the number of inspected points for sequential and nonsequential designs. The sequential designs were obtained by three rules: *MaxVar* (*darker dashed line*), *MaxWVar* (*darker solid line*), *Switch* (*darker dash-dotted line*). The nonsequential designs are: pure random sampling (*lighter solid line*) and latin hypercube sampling (*light dashed line*). For the latter designs, the data are averages of 20 replications

to $N = 50$, while the darker solid line represents the Kriging prediction at the last step of the procedure. The differences are quite evident. The *MaxVar* criterion tends to concentrate points into areas of high variation (three major details are well described), but leaves wide unexplored pieces in-between. This means that some interesting features may go undetected (see, for example, the z-shaped detail in the upper right corner). This behavior is mitigated when *MaxVar* is weighted by the inverse of the point density. Each area is sufficiently represented and no major detail is missing. The pattern from the switch criterion appears similar to the others; however, the time order of the points is different and leads to a fast increase in accuracy over the first few steps. This can be seen in Fig. 6.10, which displays the accuracy (%) of the plans derived by the three criteria. The *Switch* criterion yields an accuracy of more than 80% with six design points (only two sequential points added) and 90% with ten. The diagram also reports the accuracies of the PRS and LHS plans (the data are the averages of 20 replications), which are generally much inferior to those of the sequential plans, especially for a small-to-medium sample size. It is instructive to see how the good performance of the *Switch* criterion is generated: in the first six steps, the *MaxtInc* criterion wins, producing a rapid rise in accuracy.

6.5.2 Case 2: Roundness

The second case is taken from the technical literature (Edgeworth and Wilhelm 1999), and refers to the roundness error of a fully circular profile on a manufactured part. It is one of the few sequential inspection plans which can be found in the scientific journals of the metrological sector. The method is similar to that presented before: 360 points are measured on the circle, from 0° to 359° in steps of 1°; then design points are selected one at a time from the large dataset. Edgeworth and Wilhelm used a deterministic mechanism for next-point selection based on piecewise cubic splines interpolation.

The profile exhibits large variation and the typical wavy pattern caused by imperfect drilling operations (see the lighter line in Fig. 6.11). In particular, four lobes are observed, one of which peaks more than the others, and is the most responsible for the roundness error of the profile, which amounts to $t_{LS} = 33.7\,\mu$m. Roundness is defined as the minimum orthogonal distance between two concentric circles enclosing the actual profile. Under the mild assumption that the common center of the enveloping circles is the average of the measured points, roundness can be computed in the same way as straightness, i.e., by applying the CH method to the lighter line in Fig. 6.11.

The 50-run sequential designs obtained by the three rules are displayed in the three plots of Fig. 6.11. Here again, the *MaxVar* criterion focuses primarily on ar-

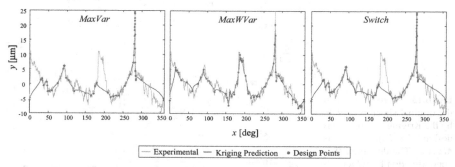

Fig. 6.11 Kriging-based sequential designs with $N = 50$ points for the straightness case study. Next-point selection rules are: *MaxVar* (*left*), *MaxWVar* (*middle*), *Switch* (*right*)

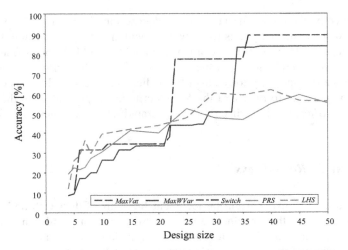

Fig. 6.12 Accuracy (%) of the computed roundness error vs. the number of inspected points for sequential and nonsequential designs. Sequential designs are obtained by three rules: *MaxVar* (*dark dashed line*), *MaxWVar* (*dark solid line*), *Switch* (*dark dash-dotted line*). Nonsequential designs are: pure random sampling (*light solid line*) and latin hypercube sampling (*light dashed line*). For these designs, the data are averages of 20 replications. Note that the *Switch* rule collapses to the *MaxVar* rule

eas of high variation. It is the only rule capable of fully capturing the most prominent lobe (thus reaching some 90% accuracy in the estimation of roundness, see Fig. 6.12), but it spends half of the point budget to do this. As a drawback, the signal approximation in other pieces of the domain is poor (Fig. 6.11). Conversely, a satisfactory reconstruction of the overall signal is provided by the *MaxWVar* rule. As the *MaxtInc* criterion is never triggered in this case, the *Switch* rule collapses to

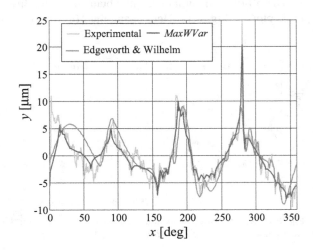

Fig. 6.13 Approximations of the original signal provided by the Kriging-based adaptive design ($N = 50$). The rule is *MaxVar*. The deterministic approximation by Edgeworth and Wilhelm is based on cubic splines (*dashed line*)

the *MaxVar* rule. The comparison with PRS and LHS is interesting. These sampling methods are not outperformed by the sequential procedures until step 21 for *MaxVar* and *Switch*, and step 26 for *MaxWVar*. This is probably due to the fact that this signal has a higher information content than that in the first case study. Therefore, more points are needed before Kriging predictions become reliable, thus making the sequential procedures more effective than the nonsequential ones. Finally, Fig. 6.13 shows how the signal reconstructed by the Kriging-based method with the *MaxWVar* rule compares with the original signal and with the approximation by Edgeworth and Wilhelm based on cubic splines for the same number (50) of collected data points. Note that the deterministic approach misses the highest peak, thus attaining only 51% accuracy.

6.6 Conclusions

The software implementing the adaptive Kriging-based inspection plans can easily be incorporated into the computer control system for the CMM and then run in real time. Thus the tool is fast, simple to automate and inexpensive. In addition to this, this chapter has demonstrated, using two applications, that Kriging plans are significantly better than those currently used in industry. We pinpoint at least two advantages of them. First, the inspection is more informed, as we allow the data to drive the sampling plan and its size. Secondly, Kriging predictions allow for a statistical (model-based) evaluation of the form error as opposed to a purely sample-based evaluation, which is plagued by a systematic downward bias. Overall, the method proposed can be regarded as a cost-effective innovation of the CMM measurement process.

An automated stopping rule, though desirable in principle, might not be good for all possible types of surface deviations; therefore, the issue of when the inspection should be stopped has not been directly addressed in the chapter. We plan to apply the method to two-dimensional form tolerances, like flatness and cylindricity. Although this extension is not difficult in theory, it may be a challenge to keep the computational time compatible with real-time operations.

References

Abrahamsen, P.: A review of Gaussian random fields and correlation function (Technical Report 917). Norwegian Computing Center, Oslo (1997)

ASME: Y14.5.1M: Mathematical definition of dimensioning and tolerancing principles. The American Society of Mechanical Engineers, New York (1994)

Badar, M.A., Raman, S., Pulat, P.S: Intelligent search-based selection of sample points for straightness and flatness estimation. J. Manufact. Sci. Eng. **125**, 263–271 (2003)

Baldi, A., Ginesu, F., Pedone, P., Romano, D.: Performance comparison of white light–optical profilometers. SEM Annual Conference & Exposition on Experimental and Applied Mechanics, St. Louis, MO, 4–7 June (2006)

Baldi, A., Pedone, P.: Caratterizzazione numerico sperimentale dei Ruotatori Geometrici nell'Interferometria in luce bianca. In: XXXIV AIAS Conf., Milan, Italy, 14–17 Sept. (2005)

Baldi A., Pedone, P., Romano, D.: Design for robustness and cost effectiveness: the case of an optical profilometer. Asian J. Qual. **7**(1), 98–111 (2006)

Buonadonna, P., Concas, F., Dionoro, G., Pedone, P., Romano, D.: Model-based sampling plans for CMM inspection of form tolerances. In: 8th AITeM Conf., Montecatini, Italy, 10–12 Sept. (2007)

Crary, S.B.: Design of computer experiments for metamodel generation. Analog Integr. Circ. Signal Proc. **32**, 7–16 (2002)

Cressie, N.A.: Statistics for Spatial Data. Wiley, New York (1993)

Cressie, N.A.: Spatial prediction and ordinary Kriging. Math. Geol. **20**(4), 407–421 (1997)

Den Hertog, D., Kleijnen, J.P.C., Siem, A.W.D.: The correct Kriging variance estimated by bootstrapping. J. Operat. Res. Soc. **57**, 400–409 (2006)

Dowling, M.M., Griffin, P.M., Tsui, K.-L., Zhou, C.: Statistical issues in geometric feature inspection using coordinate measuring machines. Technometrics **39**(1), 3–24 (1997)

Edgeworth, R., Wilhelm, R.G.: Adaptive sampling for coordinate metrology. Prec. Eng. **23**, 144–154 (1999)

Fang, K.-T, Li, R., Sudjianto A.: Design and Modeling for Computer Experiments. Chapman and Hall/CRC, London (2006)

Ghosh, B.K., Sen, P.K.: Handbook of Sequential Analysis. Marcel Dekker, New York (1991)

Giovagnoli, A., Romano, D.: Robust design via simulation experiments: a modified dual response surface approach. Qual. Reliab. Eng. Int. **24**(4), 401–416 (2008)

Goovaerts, P.: Geostatistics for Natural Resources Evaluation. Oxford Univ. Press, New York (1997)

Jones, D.R., Schonlau, M., Welch, W.J.: Efficient global optimization of expensive black-box functions. J. Glob. Opt. **13**, 455–492 (1998)

Kleijnen, J.P.C., Van Beers, W.C.M.: Application-driven sequential designs for simulation experiments: Kriging metamodelling. J. Operat. Res. Soc. **55**, 876–883 (2004)

Krige, D.G.: A statistical approach to some mine valuations and allied problems at the Witwatersrand. Master's Thesis, University of Witwatersrand, Johannesburg (1951)

Lehman, J.S., Santner, T.J, Notz, W.I.: Designing computer experiments to determine robust control variables. Stat. Sinica **14**, 571–590 (2004)

Matheron, G.: The theory of regionalized variables and its applications. Les Cahiers du Centre de Morphologie Mathematiques de Fontainbleau (France) 5 (1971)

Park, S., Fowler, J.W., Mackulak, G.T., Keats, J.B., Carlyle, W.M.M.: D-optimal sequential experiments for generating a simulation-based cycle time-throughput curve. Operat. Res. **50**(6), 981–990 (2002)

Pedone, P.: Methods for quality improvement in products and processes. PhD Thesis, University of Cagliari, Cagliari (2006)

Pedone, P., Vicario, G., Romano, D.: Kriging-based sequential inspection plans for coordinate measuring machines. Appl. Stochast. Models Bus. Ind. (in press) (2008)

Romano, D.: Sequential experiments for technological applications: some examples. XLIII Sci. Meeting of SIS (Italian Statistical Society), Turin, Italy, pp 391–402, 9–11 June (2006)

Sacks, J., Schiller, S.B., Welch, W.J.: Design for computer experiments. Technometrics **31**, 41–47 (1989a)

Sacks, J., Welch, W.J., Mitchell, T.J., Wynn, H.P.: Design and analysis of computer experiments. Stat. Sci. **4**, 409–423 (1989b)

Santner, T.J., Williams, B.J., Notz, W.I.: The Design and Analysis of Computer Experiments. Springer-Verlag, New York (2003)

Sasena, M.J., Papalambros, P., Goovaerts, P.: Exploration of metamodeling sampling criteria for constrained global optimization. Eng. Optimiz. **34**, 3, 263–278 (2002)

Schonlau, M., Welch, W.J., Jones, D.R.: Global versus local search in constrained optimization of computer models. In: N. Flournoy, N., Rosenberger, W.F., Wong, W.K. (eds) New Development and Applications in Experimental Design, vol. 34. Institute of Mathematical Statistics, Hayward, CA, pp 11–25 (1998)

Simpson, T.W., Peplinski, J.D., Koch, P.N., Allen, J.K.: Metamodels for computer-based engineering design: survey and recommendations. Eng. Comp. **17**, 129–150 (2001)

Williams, B.J., Santner, T.J., Notz, W.I.: Sequential design of computer experiments to minimize integrated response functions. Stat. Sinica **10**, 1133–1152 (2000)

Chapter 7
Product and Process Innovation by Integrating Physical and Simulation Experiments

Daniele Romano

Abstract Technical innovation in industry can massively benefit from an investigation strategy which properly combines experiments in the field with experiments on a simulation model of the product or the process. However, a methodological framework for the effective integration of the two kinds of investigation is still missing. On the one hand, simulation and lab tests are routinely used together in R&D activities of hi-tech companies, although generally not in the form of statistically designed experiments. On the other hand, *design of experiments* and *computer experiments* are sound methodologies for running experiments in physical and numerical settings, respectively, but they have practically disregarded the integration issue so far. This chapter outlines a broad approach to running a sequence of physical and simulation experiments from the viewpoint of incremental system innovation. Although the approach is still qualitative, it introduces all of the elements (system innovation, model calibration, model validation and modification, building of mechanistic models) needed to tackle a new and industrially relevant problem. The approach is demonstrated through its application to the design of an engineering system and the improvement of a production process.

7.1 Experiments and Innovation

There is undoubtedly a strong link between innovation and experiments. Innovation is the epitome of a discovery. The latter, when it does not happen by chance, comes from a deliberate process of knowledge building. The most effective of these processes is the scientific method, of which experimentation is an integral part.

In fact, the scientific method is the combination of two phases: the formulation of new hypotheses by an expert on the grounds of the currently accepted body of

Daniele Romano
Department of Mechanical Engineering, University of Cagliari
Piazza d'Armi, 1, 09123 Cagliari, Italy
e-mail: romano@dimeca.unica.it

Pasquale Erto, *Statistics for Innovation*
ISBN 978-88-470-0814-4, © Springer 2009

knowledge (deductive phase), and the verification of these hypotheses by direct observation of the system of interest (inductive phase). These two phases, repeated
sequentially, are the engine of progress in science and technology. As deductive
reasoning has presumably been a human skill for a very long time (Aristotle's foun-
dation of logic is proof of this), experiments are the missing link that previously
prevented mankind from achieving the vast progress we have experienced over the
last four centuries. From this viewpoint, the value of design of experiments (DoE),
the discipline associated with the rational management of the inductive phase, is
very clear. A statistical experiment is efficient and informative: it makes it possible
to test multiple hypotheses with controlled reliability and cost.

In the history of DoE, there is one fundamental contribution which is inspired by
the incremental nature of the scientific method. It is the approach of sequential experimentation, introduced by George Box in the 1950s (Box and Wilson 1951; Box
1999), which was particularly aimed at improving industrial systems. The optimal
setting for the system is obtained by a sequence of experimental stages with varying
objectives: ruling out inactive factors (screening); locating the region in the factor
space where the performance shows a maximum (improvement); providing accurate
prediction models for the performance by which optimal factor setting can be found
(prediction and optimization); verifying the optimal setting in the field (confirmation). The different stages represent a variable trade-off between the accuracy and
the extent (number of factors) of the analysis. While highly fractionated two-level
factorials are initially used to screen several factors, three-level response surface
designs are run in the space of the few surviving factors for prediction and optimization purposes. This results in an efficient and balanced allocation of the available budget. Notice that the statistical tools of sequential experimentation are not
particularly sophisticated nor completely new. The added value is that the methodology introduces an investigation strategy for a practical objective, i.e., achieving
system improvements at controlled costs. As this goal is highly relevant to industry,
sequential experimentation is an excellent example of how statistics can respond to
real industrial needs. Naturally, this is possible only if these needs are known. The
main elements of the strategy are reported in Table 7.1.

Table 7.1 Main features of the sequential experimentation approach

Phase	Purpose	Method	Designs
Screening	Rule out inactive factors	ANOVA	Highly fractionated 2-level factorials
Exploration	Search for a good operating region	Steepest ascent	2-level factorials
Prediction	Estimate accurate response models	Fitting of second-order response models	3-level response surface designs
Optimization	Find optimal factor setting	Analytic optimization (canonical analysis)	–
Validation	Confirm the results	Confirmatory runs near the optimum	–

7.2 Physical vs. Computer Experiments

In recent decades, another approach to investigating complex systems has become available due to the advent of computer science. The use of computer models to simulate the behavior of real systems is widespread in science and technology. As the input/output relationship of a computer model is not usually available in an explicit functional form, but rather as an I/O function which only provides the numerical output corresponding to a given input, an experimental approach is fully justified when the computing time per run is non-negligible. Thus experiments can be conducted on a system's simulator, and these represent a new source of knowledge about the system.

An important question arises: how do physical and computer investigations compare to each other? It is instructive to evaluate them on three relevant features: feasibility, cost, and fidelity to reality. If one was superior to the other in all three features there would be no point in using both! However, this is hardly ever the case: physical and computer runs generally realize a different mix of the features. Let us consider feasibility. In complex systems made up of many parts or many working mechanisms (acting at different scales, e.g., the micro or macro scales), it may be unfeasible to do physical experiments on some parts or some mechanisms due to an inability to control some inputs and/or measure some outputs, or because it is prohibitively expensive to do so. On the other hand, this can often easily be done on a simulation model. One example is measuring the internal stresses of components in mechanical design. Another two examples are found in the second case study described in Sect. 7.4, related to measurements of the electric field and the determination of trajectories of small fibers in a production process. Thus, simulation can be a substitute for a physical set-up by necessity. When both kinds of investigation are feasible, they generally provide a different trade-off between cost and fidelity. Roughly speaking, physical trials cost more, while simulations are less reliable. Setting up a physical experiment can require significantly higher expenditure than the preparation of a computer code. Moreover, the latter is often faster to run than its physical counterpart. Even in cases where computer runs are currently very time-consuming, progress in computer hardware will rapidly alleviate the problem. On the other hand, it is important to consider the risk that the computer code is unfit to reproduce the system's functioning[1].

The above discussion clearly calls for the integration of physical and computer investigations. This already occurs to some extent in advanced industrial sectors, like the aerospace, automotive, microelectronics and telecommunications industries. Here, simulations and lab trials are routinely practiced in R&D activities but not in the form of a logical sequence of experiments that are properly designed and related to each other.

---

[1] Of course, a similar risk occurs when one generalizes results obtained by experimenting on a physical prototype of the product, or on a pilot configuration of the process.

The situation is much worse in the applied statistical literature, where physical and simulation experiments are dealt with separately and hence integration is still not an issue. *Computer experiments* has been an autonomous discipline since the end of the 1980es (Sacks et al. 1989; Santner et al. 2003), but it provides a limited view of what a "computer experiment" can represent in an industrial setting. The computer model is considered expensive to run, and its output is strictly deterministic, while this is not true in several instances. In industrial research, the computer model is a tool for representing reality, and thus a subject to be improved by modifications and extensions. Real systems are inherently uncertain, since they are affected by chance and uncontrolled variables. A more consistent simulation of these systems should be a stochastic one. Yet stochastic simulation has been an important tool in many technical fields for a long time. Important examples are discrete event simulators of queueing systems used to analyze inventory, production, and telecommunication facilities where arrival and service times (of parts, phone calls, etc.) are random (Bashyam and Fu 1998). Only recently, Van Beers and Klejinen (2005) reported an application of Kriging models, the models typically adopted in computer experiments, in a stochastic simulation setting. Procedures for robust design (Taguchi and Wu 1980) developed in the context of computer experiments (Williams et al. 2000; Lehman et al. 2004) do not consider the potential of using stochastic simulation. An adaptation of the robust design logic to stochastic simulation is a very recent contribution (Giovagnoli and Romano 2008). It will be introduced in Chap. 5.

Moreover, an experienced researcher in process and product simulation is aware of the variety of ways an analysis may be carried out, all compatible with the problem at hand. Selection of the mesh replacing the real geometry of an object in finite element analysis, the choice of different competing algorithms, and decisions about convergence thresholds and the maximum number of iterations, are just some examples. Thus, a kind of "numerical uncertainty" is often present in simulation, and it is an important piece of information to consider when deciding whether the design solution is actually a good one. For an example related to nonlinear finite element simulation, see Romano and Vicario (2002).

Regarding "integration," contributions to the literature mostly address the problem of calibrating the computer model based on field data. The problem consists of considering a set of system parameters, different from the design parameters, which cannot be measured in the field but must be given a suitable value in the code (calibration). Their values are generally selected so that the simulation and field data are best matched over a target input space. This problem has been addressed with a frequentist approach (Park 1991), but more often with a Bayesian one (Bernardo et al. 1992; Craig et al. 1996; Aslett et al. 1998; Goldstein and Rougier 2003; Reese et al. 2004). However, a weakness of this approach is that it can yield an artificial match for the numerical and physical data while leaving possible model inadequacies undetected. The danger is that one has the illusion that the code is reliable and thus believes that it is safe to apply it outside the calibration region. The history of numerical studies in engineering is full of failures derived from this misunderstanding. A more sound viewpoint is to consider model calibration as only one piece of the more general problem of model validation, i.e., assessing how well the model represents reality. In fact, the computer code may be an unacceptable approximation of the physics of the system because the underlying mathematical model is inadequate or too simplified. Numerical algorithms implementing the model can also be incorrectly chosen. A good deal of work in combining calibration with validation has been done by a group of researchers at the Sandia Laboratory (Easterling 1999, 2003; Hills and Trucano 2002; Hills and Leslie 2003). In this context, Kennedy and O'Hagan (2001) introduce a fully Bayesian approach for modeling the bias between the computer model and the physical system data, and also provide for the uncertainty of predictions using a Kriging emulator of the simulation model. Their approach has been recently used by Bayarri et al. (2007) to build a framework for model validation. However, the framework is quite entangled and does not lend itself to easy application in industrial problems. It is worth mentioning an interesting attempt to integrate two simulators with different accuracies and speeds in order to generate a surrogate model which provides a convenient trade-off between accuracy and speed (Osio and Amon 1996; Qian et al. 2004). A final consideration is that, in this body of research, the role of physical observations is ancillary: they are often scarce and are not subject to design.

In this chapter, we envisage an approach for effectively integrating physical and simulation experiments. Although the approach is still broad and qualitative, it introduces the elements of a problem which is of high practical relevance: obtaining innovative findings; validating modifying the computer model in order to improve its ability to describe the real system; creating hybrid mechanistic models combining both sources of information; effectively integrating engineering and methodological expertise; and demonstrating the sequential nature of investigation. The approach is presented in Sect. 7.3. Then, two case studies, one referring to the design of a climbing robot and the other to the improvement of a manufacturing process in the textile sector, are described in Sect. 7.4. A general discussion including thoughts on future developments concludes the chapter.

7.3 An Integrated Approach

As physical and simulation experiments realize different trade-offs in feasibility, cost, and fidelity, there is an added value involved in integrating them. This section presents an approach that attempts to introduce a systemic view in this context; see Fig. 7.1. The scope is rather broad. Although the final goal is the incremental innovation of a real system, important additional objectives are improving the validation of the computer model and creating new models that incorporate the knowledge drawn from the two sources of investigation. In principle, the approach can be regarded as an extension of the conventional scientific method. The latter combines the deductive reasoning of an expert in posing new research questions with the physical observations needed to answer them, represented respectively by the middle and right branches in Fig. 7.1. The approach includes simulated observations

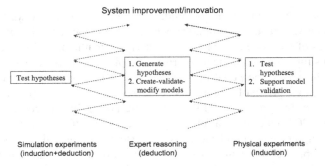

System improvement/innovation

Test hypotheses

1. Generate
   hypotheses
2. Create-validate-
   modify models

1. Test
   hypotheses
2. Support model
   validation

Simulation experiments   Expert reasoning   Physical experiments
(induction+deduction)   (deduction)   (induction)

Fig. 7.1 A schematic view of the approach to generating innovation in engineering design via integrated physical and simulation experiments

as an additional source of information (left branch in Fig. 7.1). Interestingly, these observations incorporate both induction and deduction. Induction comes from the interrogation of the code at chosen inputs, but the code itself is deduced from formalized models representing the current state of knowledge in the specific technical sector.

The approach envisages a sequential strategy managed by an expert (Fig. 7.1, middle) who, at each step, makes new conjectures based both on his technical knowledge and the freshly collected information, and then decides which source of information (physical or numerical) to use next and which experiment to run. The whole process proceeds sequentially and stops when a satisfactory level of improvement/innovation is realized. In the typical situation, where the simulator is cheaper (faster) and the physical set-up is more reliable, it is sensible to use simulation experiments to explore the space of the design variables in depth in order to get innovative findings, and to use a moderate amount of costly physical trials to verify the findings. This is particularly wise in product development projects, where the cost of prototypes is a major concern. In such a case, it is preferable to search for interesting product configurations using numerical experiments, and to only build the physical prototype afterwards in order to prevent practical constraints like time, cost, and manufacturability from limiting the designer's creativity at a very early stage. An example is shown in the application presented in Sect. 7.4.1. If findings obtained by simulation are not confirmed in the field, the computer code should be revised accordingly. The rationale is that the comparison of physical and numerical results should allow not only for model calibration but should also provide the expert with some hints on how to modify the mathematical model of the simulator in order to improve its ability to reproduce the system. This of course implies a high level of technical expertise regarding the physics of the system and the simulation models, and it generally requires qualified teamwork. Thus physical experiments can play different roles. They are used not only to test expert hypotheses but also to verify numerical results and to support dynamic code review.

The key element is the expert–computer interaction. Exploration brings unexpected results to the expert who, in turn, activates creative thinking from them. This synergistic loop makes the path towards innovation easier and systematic, and, in some cases, can breed radical innovation. Successful applications to the design of a force transducer can be found in Barbato et al. (1997) and Baldi et al. (2006). In both cases, innovative design solutions that gave results well beyond the initial designers' expectations were achieved. Another outcome of the approach can be a hybrid model of the product or the process that integrates the information drawn from physical and computer experiments. An example of such model is found in the application described in Sect. 7.4.2.

7.4 Applications to Product and Process Development

Two case studies are now presented to illustrate the integrated approach. They refer to product design and process improvement, respectively. Schematic views of the sequence of steps involved in the integrated approach are provided in Fig. 7.2. Each experimental step, either physical or numerical, is followed by a step representing the expert's deductive activity (data analysis, conjectures, design of new experiments), on which the next experimental step is based. The description follows the stylized structure of the sequential procedure.

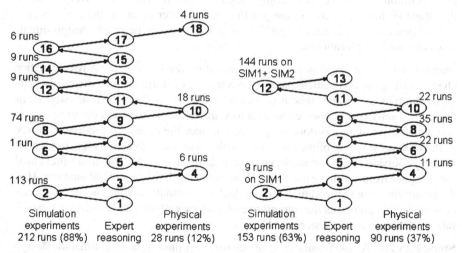

Fig. 7.2 Application of the approach to new product design (*left*) and to process improvement (*right*)

7.4.1 Product Innovation: The Climbing Robot

This application concerns the design of a pneumatic climbing robot, realized at the Dept. of Mechanical Engineering of the University of Cagliari, Italy (Manuello et al. 2003; Atzori 2003). The device (Fig. 7.3) must be capable of reaching the tops of vertical structures—posts, trees, bridge cables—while carrying equipment that will allow it make diagnoses and possibly operate on the structures. In order to cling onto the post, the robot exploits a passive stop mechanism using locking rings. The elimination of a dedicated actuator for clinging makes the robot lighter. The design objective is to make the upward motion of the robot fast and stable regardless of the surface conditions of the post. This is a robust design study where the friction between the post material and the locking rings is an external noise factor. The design strategy is made up of 18 steps, which are now concisely described.

Steps 1 and 2. A set of 21 factors, including geometrical and mechanical variables, are selected for a vast exploration campaign to be performed on the computer. The simulation model is built using a commercial package for mechanical design, *Working Model*. This exploration would never have been made with physical experiments as it would have required a dozen different prototypes. Six important factors are identified via two screening experiments (a 32-run Plackett–Burman design and a 32-run 2^{14-9} fractional factorial) and a response surface estimated (Fig. 7.4) with a 49-run Box–Behnken design (Myers and Montgomery 2002). The response is an indicator of the robot's ability to climb with the zero level discriminating between climbing (positive) and falling (negative). An important discovery made at this stage is that, by changing the position of its center of mass, the robot is also able to descend in a controlled fashion, which is beyond the initial design intent. Thus computer exploration has promoted innovation.

Steps 3 to 6. A physical prototype is realized on the basis of these results. After characterizing the mechanical parameters via a set of static measurements, its ability to climb up a Plexiglas tube is proven in a small set of dynamic tests (step 4). In these tests, only two easy-to-change factors are varied. However, a direct comparison with the simulation is done only at step 6, since the measured pre-loading force of the transverse spring of the prototype, which was found to be the most influential factor, is outside the range explored in step 2. In fact, a cost constraint prevented the designer from adopting a configuration in line with the numerical analysis. Had the prototype been realized first, a great deal of information would never have been collected. This pinpoints the importance of starting with extensive computer exploration, almost free of practical constraints.

Steps 7 and 8. As the feasibility analysis for the prototype has modified the design region, a new cycle of computer experiments is needed to search for a region where the robot is capable of climbing with a fast and regular motion on virtually all post surfaces. This is achieved with a sequence of two experiments, a six-factor Box–Behnken design, and a four-factor central composite design. In the first experiment, three factors do not belong to the influential set identified after the first computer

Fig. 7.3 The prototype of the robotic device in the lab set-up

Fig. 7.4 The surface response marking the limit between climbing and falling conditions in the space of the three most influential factors

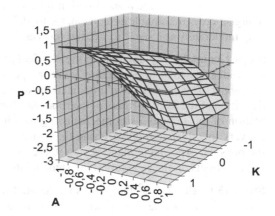

exploration. These are easy-to-change factors that replace the same number of factors that have feasible fixed values in the physical prototype. The experiment turns out to be extremely instructive for the expert, as it reveals several robot behaviors (monotonic climb, nonmonotonic climb, climb and then fall, no move, no move and then fall, descend in control, fall). Moreover, it provides evidence that when the angle between the locking rings and the post direction approaches 90°, the robot is robust to variations in the friction conditions in the ring–post contact. The second experiment is aimed at identifying a region where the robot can always climb.

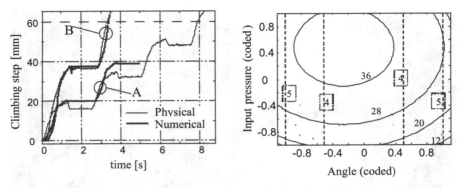

Fig. 7.5 *Left*: physical system vs. simulation: before model validation (*A*); after model validation (*B*). *Right*: Contour lines of the mean (*solid*, mm) and the variance (*dashed*, mm^2) of the climbing step length for the optimized design configuration

Steps 9 to 12. A confirmatory experiment (a 2^3 factorial plus center point with two replications) is run in the lab using the most important factors from the previous analysis (steps 9 and 10). This confirms the findings regarding robustness. Since the design region was slightly resized for feasibility reasons, an identical computer experiment is run in the resized region for model validation (steps 11 and 12). Although there is reasonable agreement, the model is shown to be inadequate (although only moderately so), see Fig. 7.5, left (curves A).

Steps 13 and 14. By contrasting the physical and simulation results, the expert is in an ideal position to conjecture about the possible causes of this inadequacy of the model. This leads to two modifications of the computer code: a more accurate definition of the axial stiffness of the robot as a function of its elongation, and a simulation of the *stick and slip* mechanism governing the contact of the two surfaces for very low relative motion. Using the revised code for a new computer replica of the physical experiment made at step 9, the agreement is improved. Now the three largest effects on the robot climbing step and the signs of the others coincide.

Steps 15 to 18. The 2^3 factorial is augmented with center and axial runs (step 15 and 16) for robust design optimization (step 17, Fig. 7.5, right). The optimum is eventually confirmed by one lab test (step 18, Fig. 7.5, left (curve B)). The total experimental cost was quite reasonable, as demonstrated by the allotment of runs in the two settings: only 12% physical runs and 88% computer runs.

7.4.2 Process Innovation: The Flocking Process

The second application was developed in an industrial research project. A medium-sized Italian textile firm produces flocked yarn; after weaving, this fabric is used in a wide range of technical applications. Typical end-products are coverings for seats and other components in car interiors. The yarn is formed from finely cut polyamide fibers (flock) applied to an adhesive coated carrier thread. The flock confers a smooth texture and other important technical characteristics to the surface, like water repellency, resistance to abrasion and to light. The project was motivated by the need to enhance product quality. One of the most critical quality characteristics is the "title," related to the flock density on the yarn. It is measured in dtex (1 dtex is the mass, in grams, per 10^4 m yarn length). Out-of-spec conditions for the title were experienced all too often in terms of both low average title and excessive scatter.

The basic flocking process is illustrated in Fig. 7.6. While an array of parallel coated threads passes in-between the electrodes of a high-voltage capacitor (the actual production lines contain three consecutive capacitors), the flock fibers are fed to a conveyer belt which takes them into the electric field generated by the capacitor. The electrostatic force aligns the flock fibers (which have been conditioned to have sufficient conductivity) and pushes them upwards until they hit a thread and attach to it. As each thread also spins on its axis, the flock implants itself all around it. The unsuccessful fibers are recovered and brought back to the flock feeder.

The integrated approach is described according in Fig. 7.2. Several sources of information are used. Physical experiments are conducted in a pilot plant, in production lines and in a lab set-up. Two simulation models are developed and integrated in order to predict the basic mechanisms involved in the process: the electric field generated by the electrodes and the rising path of the flock between the electrodes.

An initial analysis of a huge mass of historical process data measured during production runs revealed—quite surprisingly—weak correlations between process variables and the title. Even the electrical variables (electrode voltages), the driv-

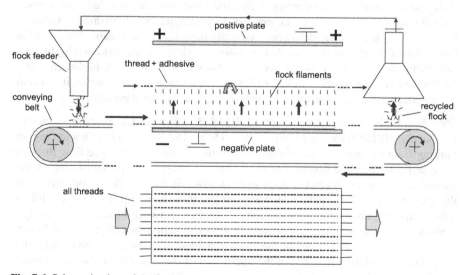

Fig. 7.6 Schematic view of the flocking process

Table 7.2 List of the selected process parameters, classified by type

Selected process parameters (units)			
Electrical	Mechanical	Geometrical	Environmental
Voltage of lower plate, V^- (kV)	Thread speed, v_t (m/min)	Flock size, F (qualitative)	Absolute humidity, H_a (g/kg of dry air)
Voltage of upper plate, V^+ (kV)	Thread pull force, P (N)	Groove of glue shaft, G (qualitative)	Relative humidity, H_r (%)
–	Conveyer speed, v_c (m/min)	–	–

ing force of flock motion, appeared nearly inert in this respect. It was decided that new process data would be collected under strictly controlled conditions for two consecutive days. This time, the expected input–output correlations clearly stood out. A close enquiry revealed that operators measured parameters at the beginning of their shift but omitted to track the manual adjustments they made to parameters during the shift. This was a decisive discovery that convinced the company to embark on a structured experimental investigation. Technicians were then asked to list the most important parameters; see Table 7.2. Notice that, since the voltage of the lower electrode is always negative, factor V^- is taken to be the absolute value of the voltage. The 13 steps of the integrated approach are described below.

Step 1. Conjecture 1. As the threads are kept at ground potential (0 kV), the typically adopted antisymmetric voltage setting of the electrodes, $V^+ = -(-V^-)$, generates an approximately uniform electric field with straight field lines in the flocking chamber. The consequence is that the flocks that fail to hit the threads during their rise are push towards the upper electrode and must be recycled. This causes a substantial efficiency loss in the process. In principle, by establishing symmetry, $V^+ = (-V^-)$, the electric field could virtually push the flock towards the thread at both sides of the chamber. Since this operating condition was considered highly unsafe by the technicians, it would never have been tried in the field. In this case, the finite element electrostatic simulator (SIM 1) turns out to be very useful. It has also a high fidelity level, as it solves a system of linear differential equations.

Step 2. Simulation experiment 1 on SIM 1. A 3^2 full factorial design with factors V^- (levels: 10/15/20 kV) and V^+ (levels: −15/0/+15 kV) is run over a wide region ranging from the antisymmetric condition to the symmetric one. The flow lines of the electric field in the first two plots of Fig. 7.7 confirm the previous conjecture. Electrical symmetry forces the flock to point towards the threads in both halves of the chamber (Fig. 7.7, center). However, at the threads, the electric field exhibits a large discontinuity which might cause discharges between close flock particles. Anyway, field discontinuity can be mitigated by increasing the upper voltage. When it is set to zero, the flow lines of the electric field seem to be acceptable while the advantage of realizing flocking in the upper part of the chamber is also maintained (see Fig. 7.7, right).

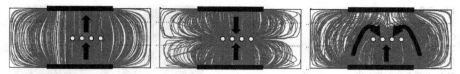

Fig. 7.7 Flow lines of the electric field in three voltage settings. *Left*: $V^- = -15$ kV, $V^+ = 15$ kV; *center*: $V^- = -15$, $V^+ = -15$; *right*: $V^- = -15$ kV, $V^+ = 0$ kV

Step 3. Conjecture 2. The innovative process setting with null voltage at the upper electrode is worth testing in the field. However, since it is still too far from the current setting, it is not safe enough to be tried on production lines. Therefore, it was decided that an old pilot plant would be reconditioned, which, although it is a simplified and smaller version of the real plant, has the advantage that the flocks rising in-between the electrodes can be seen. This is extremely useful in order to check the flock trajectories induced by the electric field.

Step 4. Physical experiment 1 in the pilot plant. An 11-run 2^{7-4} fractional factorial of resolution III plus one quasi center point replicated three times is run for screening purpose. The factors are: V^- (levels: 24/33.5/43 kV), V^+ (levels: grounded/disconnected), thread speed (levels: 7.75/9.3/10.85 m/min), thread pull force (levels: 9/−10.75/12.5 N), ratio between thread speed and conveyer speed (levels: 1.2/1.5/1.8), flock size (levels: small/large), absolute humidity (levels: 12.5/14.5/16.5 g/kg). On each of the four yarns, five measurements of the title are taken by weighing a fixed yarn length. The analyzed responses are the title average and scatter, computed over both the yarns and the repeated measurements. The voltage of the lower electrode is by far the dominant factor for the average title, while, contrary to the technicians' belief, the effect of humidity is not significant (Fig. 7.8, left). The title is significantly higher than the value obtained with the typical production setting (+64% on average), but the scatter is also excessive and it increases with time, indicating that the process is not stable (Fig. 7.8, right). Only the effect of flock size appears significant for scatter. Pictures taken during operations confirm

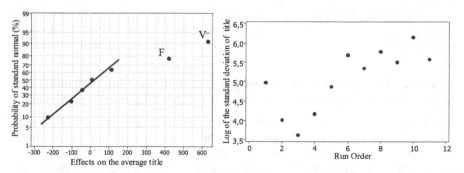

Fig. 7.8 Outcome of physical experiment 1 in the pilot plant. *Left*: normal probability plot of the effects on the average title; *right*: scatterplot of the title dispersion

Fig. 7.9 Efficient but turbulent flocking during physical experiment 1 (run with $V^- = -43\,\text{kV}$, $V^+ = 0\,\text{kV}$). Flock rising from the sides rains down the threads

that flocking now occurs in both halves of the chamber. In Fig. 7.9, one can see the curved flock trajectories pointing from the sides to the center, consistent with the field lines observed in the simulation. However, it also shows rather chaotic flocking operations, with the flock particles colliding with each other. Discharges were often observed. The effect of flock size on scatter is probably due to the fact that larger particles are more likely to collide.

Step 5. Conjecture 3. Conditions for a much higher process yield have been obtained but operations are still unsafe. A more conservative voltage for the upper electrode should stabilize the process while preserving the enhanced yield. A decision is made not to experiment on the production line yet.

Step 6. Physical experiments 2 and 3 in the pilot plant. Another 11-run 2^{7-4} fraction is run, with the exception here that V^+ is a quantitative factor whose levels are positive ($+18/+24/+30\,\text{kV}$), albeit inferior to the corresponding levels of V^-. The good outcomes of this experiment suggested that a foldover experiment should be run in order to break aliasing (see Myers and Montgomery (2002) for details on fold-over designs) between the main effects and two-factor interactions. Results indicate that the process is now stable, and conditions that increase the average title and decrease its scatter can be found, as shown in Fig. 7.10. Dot plots of experimental responses show that results are always better than those obtained with the typical voltage setting used in production. The process is governed by the electrode voltages. In particular, V^- controls the average and V^+ the variability. A high V^- and a positive but low V^+ increase the average title and reduce its variability. Higher thread speed yields a lower title, as expected. However, this effect is easily compensated for by increasing V^-. This finding is the key to increasing productivity.

Step 7. Conjecture 4. As a satisfactory process improvement has been realized in the pilot plant, it is important to assess whether it can be transferred to production.

Step 8. Physical experiments 4 and 5 on the production line. A 20-run 2^{7-3} fraction of resolution IV $+2$ quasi center points replicated twice is run. Factors are: V^- (levels: 15/20/25 kV), V^+ (levels: 5/10/15 kV), thread speed (levels: 14/17.5/21 m/min), thread pull force (levels: 8/9/10 N), relative flock density (levels: 30/50/70%), size of grooves (levels: small/large), relative humidity (levels: 50/60/70%). A small flock

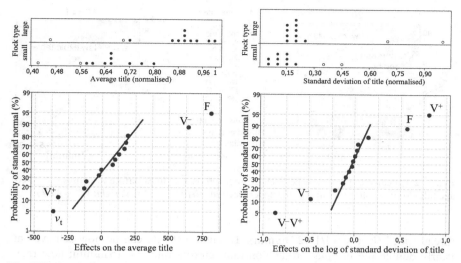

Fig. 7.10 Outcomes of physical experiments 2 + 3 in the pilot plant. Responses are title average and dispersion. *Top*: dot plots of responses; *bottom*: normal probability plot of responses

size was used. Since the thread traction system is more efficient on the production line, the speed can be increased in order to verify whether higher production rates are feasible, as suggested by the experiment in the pilot plant. Five title measurements were taken on six yarns belonging to an 80-thread array (the innermost pair and the two outermost pairs). Results generally confirmed the expectation of process improvement in terms of increased title (see Fig. 7.11). Title variability was also within the admissible range. The good results prompted a search for the optimal process conditions, and a response surface design (a 15-run Box–Behnken design with three center points) was run, restricting attention to the most influential and easy-to-change process variables: V^- (levels: 20/25/30 kV), V^+ (levels: 4/12/20 kV), and

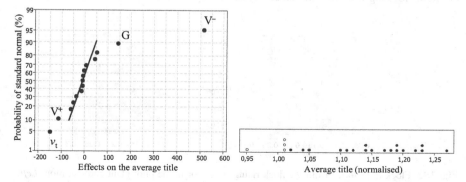

Fig. 7.11 Outcome of physical experiment 4 on the production line. Response is the average title. *Left*: normal probability plot of the effects; *right*: dot plot of the response

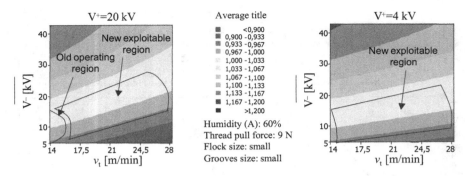

Fig. 7.12 Outcome of physical experiment 5 on the production line. Surface responses (i.e., average title) for two voltage settings of the upper electrode (*left*: $V^+ = 20\,\mathrm{kV}$, *right*: $V^+ = 4\,\mathrm{kV}$)

thread speed (levels: 14/19/24 m/min). The nominal title can be now achieved at nearly double the original production rate (current rate: 14–15 m/min; new rate: 24 m/min or more) by proper tuning of the electrode voltages. As an example, new exploitable operating regions are spotted in the response surfaces plotted in Fig. 7.12 for two different settings of the voltage of the upper electrode.

Step 9. Conjecture 5. The creation of a full mechanistic model of the flocking process would be very useful for design purpose. This requires the development of a dynamic simulator of the flock rising (SIM2) to be integrated with the electric simulator. The idea is that, for a given process configuration, the electric field obtained by SIM1 is input to SIM2, which, upon computing the flock trajectories for all starting positions on the conveyer belt, verifies whether the flock hits a thread or not. SIM2 was developed in MATLAB by integrating the laws of motion for a flock particle under the influence of the electric field, gravity and air resistance.

A typical output of the program is depicted in Fig. 7.13, which shows flock trajectories for two different voltage settings of the pilot plant set-up. The plots clearly confirm the finding that decreasing V^+ and increasing V^- (right) leads to much more efficient flocking than that achieved in the setting adopted formerly. However, this

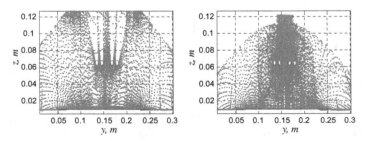

Fig. 7.13 Flock trajectories from the flock rising simulator in the pilot plant configuration. *Left*: $V^+ = 30\,\mathrm{kV}$, $V^- = -24\,\mathrm{kV}$, *right*: $V^+ = 0\,\mathrm{kV}$, $V^- = -33.5\,\mathrm{kV}$

information is not sufficient to build a full mechanistic model, because SIM2 only helps to determine whether a flock will hit the thread. It does not provide the rising rate of the flocks packed on the conveyer belt. An ad hoc experiment was designed that exploited a lab set-up called *electrostatic jump*, which was routinely used to test flock conductivity. In this test, a fixed amount of flock is lifted from a plate under the pull of a known electric field. The time needed to empty the plate is an indicator of flock conductivity. By discontinuing the electric field at different points in time and weighing the quantity of flock remaining on the plate at each step, it is easy to estimate the rising rate as $\mathrm{d}n/\mathrm{d}t(t=t_i) \sim m_{\mathrm{f}}^{-1}\Delta m_i/\Delta t_i$, where n is the number of lifted flocks, Δm_i is the mass of lifted flocks in the time lag $\Delta t_i = t_i - t_{i-1}$, and m_{f} is the mass of a single particle.

Step 10. Physical experiment 6 in the lab. A 22-run $3 \times 2 \times 3$ full factorial design with two quasi center points replicated twice is run in two blocks. The factors are: the electric field on the plate (levels: 3/7.15/5.3 Kv/cm), the flock type (small, large), the initial amount of flock on the plate ($0.25/0.50/0.75 \times 10^{-3}$ kg). As shown in Fig. 7.14, the results support a prey–predator differential model for flock rising, where the rising rate is proportional to the number of flocks available on the plate

$$\frac{\mathrm{d}n}{\mathrm{d}t} = a(n_0 - n), \quad n(0) = 0 \tag{7.1}$$

Here $n(t)$ is the number of flocks lifted up to time t, n_0 is the number of flocks initially present on the plate, and a is the diffusion parameter. This parameter characterizes the process and depends on both the flock type and the electric field on the starting plate, as is apparent in Fig. 7.14. The solution to Eq. 7.1, $n(t) = n_0(1 - \mathrm{e}^{-at})$, impies that flocking becomes less and less efficient over time, i.e., with as the conveyer belt becomes increasingly empty. A practical consequence of this is that the more efficient flocking could be exploited to simplify the process by eliminating one or two pairs of electrodes in the current configuration (which has three).

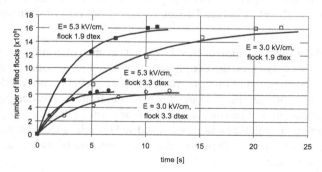

Fig. 7.14 Outcome of the lab experiments on flock rising rate. *Solid lines* follow the exponential model—solution to Eq. 7.1; *squares and circles* refer to experimental data

Step 11. Conjecture 6. There is now enough information for a mechanistic model for the title to be deduced. Consider the contribution dn_i from flock implanting on yarn i, $i = 1$ to N, during the time interval dt around time t, that comes from the elementary area dA around the point (x, y) on the conveyer belt:

$$dn_i = dn \cdot S(x, y) \tag{7.2}$$

where $S(x, y)$ is a binary function that is one if the flock rising from location (x, y) hits the thread i and zero otherwise. Note that function S is the outcome of the integrated simulator SIM1+SIM2. Using Eq. 7.1, where n_0 must be replaced by the initial flock quantity on the element of area dA, i.e., $n_0 L^{-1} W^{-1} dx dy$ for a uniform flock distribution on the conveyer belt, Eq. 7.2 can be rewritten as:

$$dn_i = a n_0 L^{-1} W^{-1} e^{-at} S(x, y) dx dy dt \tag{7.3}$$

Using process stationarity ($t = x v_c^{-1}$) and integrating (7.3) over the whole conveyor belt ($0 \leq x \leq L$, $0 \leq y \leq W$) and over the time interval T needed to allow 10^4 m of yarn to pass into the chamber ($T = 10^4 v_t^{-1}$), one obtains the model for the title on each yarn i of the array:

$$\text{title}_i = 10^4 m_t + n_i m_f \tag{7.4}$$

where m_t is the mass of one meter of thread.

Conjecture 7. The flock motion simulator requires validation and calibration. For validation, the air resistance opposing the motion can be modeled by either aerodynamic resistance or viscous friction. In each case there is only one parameter to be tuned: the drag coefficient for aerodynamic resistance and the viscosity coefficient for viscous friction. Another calibration parameter is flock charge.

Step 12. Simulation experiments 2 and 3 on SIM1+SIM2. For each modeling choice a 72-run factorial experiment is run where two factors at three levels (flock charge and drag coefficient for the aerodynamic resistance model; flock charge and viscosity coefficient for the viscous friction model) are fully crossed with the eight different electrical settings tested in the physical experiments (1, 2 and 3) performed in the pilot plant. The plots in Fig. 7.13 are based on two runs of simulation experiment 2.

Step 13. Validation/calibration of SIM2. Validation and calibration are performed in a combined way. Each of the two models (aerodynamic resistance and viscous friction) will be calibrated in order to minimize the bias between the title obtained via the mechanistic model (Eq. 7.4) and that resulting from the physical experiments (1, 2 and 3) in the pilot plant (on each of the four yarns in that set-up). The model with the minimum overall bias will be validated and the relevant calibration adopted. This activity is still ongoing. Then the validated model will be used as an important tool for process design.

7.5 Discussion and Future Developments

It is instructive to assess whether the case studies support the statement that the integrated use of physical and simulation experiments is better than using physical experiments alone. In the first case study, an unexpected design innovation, that the robot can descend steadily without any active mechanism by simply changing its mass distribution, came from the initial computer exploration covering 21 factors. The location of the center of mass would not have been included in the factor set if an expensive physical experiment had been done instead. The optimized robot can climb at a speed that is seven times higher than that of the initial design configuration, and it is also robust, i.e., it can climb on slippery posts too. These results were obtained with 28 physical trials, only 12% of the total, made with just one prototype, and built according to the indications from the computer exploration. More than two hundred simulations were run on a reliable engineering package that required only a small amount of computing time. This is the best scenario for generating innovation: expensive physical runs and cheap simulations. Switching to the second application, the total number of runs was practically the same (243), but more than one third of these were physical. Experimenting in existing facilities (pilot plant, production lines, a lab set-up that was already used for compulsory tests on raw materials) meant that physical trials were not very expensive while computer runs (flock trajectories were obtained by the integration of a nonlinear dynamic system) were not that cheap. This is a different situation which is more likely to produce an improvement. Indeed, the improvement in the process was remarkable. The company learnt that the process is ruled by its electrical parameters, and that these can be used in a flexible way, i.e., the desired title can be obtained over a much wider specification range than the one used in the past. This knowledge can be exploited in different ways. First, productivity can be massively increased. Second, the process can be simplified. The nominal title can be obtained using only one pair of electrodes, as was the case in the pilot plant. Third, more efficient flocking means less recycled flock, and this in turn leads to better yarn quality, since flock properties deteriorate with each passage through the chamber. Finally, the ability to achieve a significantly higher title than before resulted in the generation of new ideas for products in previously unexplored market sectors, like clothing and cleaning. Therefore, process flexibility opens the door to product innovation. Could this also have been achieved using only physical experiments? Operating conditions considered unsafe would undoubtedly have been ignored in field tests. Likewise, the hybrid mechanistic model of the process is a unique outcome of the integrated approach.

It is important to emphasize that the approach presented here has a loose structure. It does not go much beyond the definition of the elements of the problem. The decisions that determined the sequential paths in Fig. 7.2 were mostly subjective. This makes the success of the investigation critically dependent on the presence of a team of experts that combine, and possibly share, skills in statistics, engineering, and computer science. Admittedly, this is more natural for a high-level strategy where the objective could change as a result of feedback produced by the evolving knowledge. Commenting on the sequential experimentation approach, George Box warned: "The reader should notice the degree to which informed human judgment decides the final outcome" (Box et al. 1978, p. 537). However, there is the need for quantitative methods that support the decisions involved in the approach. Different decision levels need to be handled. High-level decisions include: whether to stop or continue the whole investigation; whether to conduct the next experiment on the physical system or on its simulator; and defining the purpose of the experiment (exploration, improvement, confirmation, model validation). Intermediate-level decisions include the location of the experimental region and the run size. A low-level decision is the choice of the experimental design. Decisions should balance a number of criteria, including measures of expected improvement, the cost of runs, and the fidelity level. The author is currently working on this topic.

On a few occasions in the history of design of experiments, comprehensive methodologies addressing actual industrial needs have been devised. Two of the most significant are sequential experimentation and robust design (Taguchi and Wu 1980). The integration of physical and computer experiments is expected to be another of these methodologies. It should be an asset for industries that are willing to search for systematic technical innovations in our knowledge-based society.

Acknowledgements The author wishes to thank Prof. A. Manuello, Dr. M. Ruggiu, Dr. P. Pedone, Dr. M. Sandigliano and Dr. S. Masala for their valuable contributions to the development of the two case studies reported in this chapter.

References

Aslett, R., Buck, R.J., Duvall, S.G., Sacks J., Welch W.J.: Circuit optimization via sequential computer experiments: design of a output buffer. Appl. Stat. **47**, 31–48 (1998)

Atzori, M.: Ottimizzazione di un dispositivo rampicante su pali. Master's thesis, University of Cagliari, Cagliari (2003)

Baldi, A., Pedone, P., Romano, D.: Design for robustness and cost effectiveness: the case of an optical profilometer. Asian J. Qual. **7**(1), 98–111 (2006)

Barbato, G., Romano, D., Zompì, A., Levi, R.: Sperimentazione Numerica per la Progettazione di Elementi Dinamometrici a Colonna. 26th AIAS Conf., Catania, Italy, pp. 327–334, 2–6 Sept. (1997)

Bashyam, S., Fu, M.C.: Optimization of (s, S) inventory systems with random lead times and a service level constraint. Manag. Sci. **44**(12), 243–256 (1998)

Bayarri, M.J., Berger, J.O., Paulo, R., Sacks, J., Cafeo, J.A., Cavendish, J., Lin, C.-H., Tu, J.: A framework for validation of computer models. Technometrics **49**(2), 138–154 (2007)

Bernardo, M.C., Buck, R.J., Liu, L., Nazaret, W.A., Sacks, J., Welch, W.J.: Integrated circuit design optimization using a sequential strategy. IEEE Trans. Comput.-Aided Des. **11**, 361–372 (1992)

Box, G.E.P., Wilson, K.B.: On the experimental attainment of optimum conditions, J. R. Stat. Soc. B **13**, 1–45 (1951)

Box, G.E.P., Hunter, W.G., Hunter, J.S.: Statistics for Experimenters. Wiley, New York (1978)

Box, G.E.P.: Statistics as a catalyst to learning by scientific method, Part II—a discussion (with discussion). J. Qual. Technol. **31**, 16–29 (1999)

Craig, P.S., Goldstein, M., Seheult, A.H., Smith, J.A.: Bayes linear strategies for matching hydrocarbon reservoir history and discussion. In: Bernardo, J.M. et al. (eds.): Bayesian Statistics 5. Oxford University Press, Oxford, pp. 69–95 (1996)

Easterling, R.G.: A framework for model validation (Technical Report SAND99-0301C). Sandia National Laboratories, Albuquerque, NM (1999)

Easterling, R.G. Statistical foundations for model validation: two papers (SAND2003-0287). Sandia National Laboratories, Albuquerque, NM (2003)

Giovagnoli, A., Romano, D.: Robust design via simulation experiments: a modified dual response surface approach. Qual. Reliab. Eng. Int. **24**(4), 401–416 (2008)

Goldstein, M., Rougier, J.C.: Calibrated Bayesian forecasting using large computer simulators (technical report). Statistics and Probability Group, University of Durham, Durham (2003)

Hills, R.G., Trucano, T.G.: Statistical validation of engineering and scientific models: a maximum likelihood based metric (SAND2001-1783). Sandia National Laboratories, Albuquerque, NM (2002)

Hills, R.G., Leslie, I.: Statistical validation of engineering and scientific models: validation experiments to application (SAND2003-0706). Sandia National Laboratories, Albuquerque, NM (2003)

Kennedy, M.C., O'Hagan, A.: Bayesian calibration of computer models. J. R. Stat. Soc. B **63**(3), 425–464 (2001)

Lehman, J.S., Santner, T.J., Notz, W.I.: Designing computer experiments to determine robust control variables. Stat. Sinica **14**, 571–590 (2004)

Manuello, A., Romano, D., Ruggiu, M.: Development of a pneumatic climbing robot by computer experiments. Ceccarelli, M. (ed.): Proc. 12th Int. Workshop on Robotics in Alpe-Adria-Danube Region, Cassino, Italy, 7–10 May (2003); available on CD-ROM

Myers, R.H., Montgomery, D.C.: Response Surface Methodology, 2nd edn. Wiley, New York (2002)

Osio, I.C., Amon, C.H.: An engineering design methodology with multistage Bayesian surrogates and optimal sampling. Res. Eng. Des. **8**, 189–206 (1996)

Park, J.S.: Tuning complex computer codes to data and optimal designs. Ph.D. thesis, University of Illinois, Urbana-Champaign, IL (1991)

Qian, Z., Seepersad, C.C., Joseph, V.R., Allen, J.K., Wu, C.F.J.: Building surrogate models based on detailed and approximate simulations (ASME Paper no. DETC2004/DAC-57486). In: Chen, W. (Ed.): ASME 30th Conf. of Design Automation, Salt Lake City, USA. ASME, New York (2004)

Reese, C.S., Wilson, A.G., Hamada, M., Martz, H.F., Ryan, K.J.: Integrated analysis of computer and physical experiments. Technometrics **46**(2), 153–164 (2004)

Romano, D., Vicario, G.: Reliable estimation in computer experiments on finite element codes. Qual. Eng. **14**(2), 195–204 (2001–2002)

Sacks, J., Welch, W.J., Mitchell, T.J., Wynn, H.P.: Design and analysis of computer experiments. Stat. Sci. **4**, 409–435 (1989)

Santner, T.J., Williams, B.J., Notz, W.I.: The Design and Analysis of Computer Experiments. Springer-Verlag, New York (2003)

Taguchi, G., Wu, Y.: Introduction to Off-Line Quality Control. Central Japan Quality Control Association, Nagoya (available from American Supplier Institute, Romulus, MI, USA) (1980)

Van Beers, W.C.M., Kleijnen, J.P.C.: Customized sequential designs for random simulation experiments: Kriging metamodeling and bootstrapping (Discussion Paper no. 55). Tilburg University, Tilburg (2005)

Williams, B.J., Santner, T.J., Notz, W.I.: Sequential design of computer experiments to minimize integrated response functions. Stat. Sinica **10**, 1133–1152 (2000)

Chapter 8
Continuous Innovation
of the Quality Control of Remote Sensing Data
for Territory Management

Elisabetta Carfagna and Johnny Marzialetti

Abstract This chapter deals with the problem of assessing the quality of land-cover databases, since only high-quality products are useful for gaining knowledge about and managing territory. After a brief analysis of the main aspects of quality control and validation of land-cover databases, the main concepts of statistical quality control methods are recalled in order to show how some quality control procedures for land-cover databases can be formalized and improved by taking advantage of statistical quality control methods. Then, sequential and two-step adaptive procedures with various quality indices are proposed that continuously improve the quality of land-cover databases during the production process, in order to satisfy the user's needs.

8.1 Land-Cover Databases

Several land-cover databases have been produced over at least the past few decades that use aerial photos or high-, medium- and coarse-resolution satellite data. These types of databases have become the most important instrument for gaining knowledge about and managing the territory at various scales, depending the kind of utilization required, ranging from a local to a global scale. Since these databases give information about land cover and land cover dynamics (when used repeatedly over time), they are the main source of information for very different activities, such as simulating the impact of alternative policies in a region in order to improve the utilization of the territory, estimating soil pollution and planning incentives for farm-

Elisabetta Carfagna
Department of Statistical Sciences, University of Bologna
e-mail: elisabetta.carfagna@unibo.it

Johnny Marzialetti
Department of Statistical Sciences, University of Bologna
e-mail: johnny.marzialetti@unibo.it

Pasquale Erto, *Statistics for Innovation*
ISBN 978-88-470-0814-4, © Springer 2009

ers in order to reduce their pollution of the soil, stratifying the territory, estimating parameters of models for research into global changes, and so on.

Public administrations and the scientific community are the main users of land-cover databases, which are basically digital maps. They are produced by the photointerpretation of images on the screen or by the semiautomatic classification of a set of measures of the electromagnetic radiation reflected by a unit area of the Earth's surface. These unit areas are called pixels and they can range in size from less than 1 m to 5 km.

Various kinds of classifiers have been developed for semiautomatic classification; some of them perform the classification pixel-by-pixel (e.g., maximum likelihood classifier), some others classify contiguous groups of pixels (e.g., parallelepiped classification), while some perform supervised classification and some others perform unsupervised classification.

Supervised classification, as well as photointerpretation of remote sensing data, is performed according to a land-cover legend that is defined in advance. In the legend, each class (or label) represents a land-cover type. In semiautomatic classification, more classes than the ones foreseen by the legend are often used in order to catch the variability inside the training set; some classes are then aggregated.

The result of photointerpretation or semiautomatic classification is a database created in a geographic information system (GIS) whose basic elements are pixels (in a raster approach) or polygons (in a vectorial approach) of specific land-cover types. The raster approach is commonly used when pixel-by-pixel semiautomatic classification is performed. The GIS allows many kinds of operations to be performed on the polygons, such as the division of a polygon into two pieces, merging, overlaying different polygons, and so on.

A land-cover database is useful if its quality is evaluated and is high and so the database can be considered reliable. Sometimes, reliability is confused with the scale of the product: the more detailed the scale, the higher the reliability.

The scale of remote sensing data represents only the level of detail of the basic material and cannot be considered the quality of the land-cover database. Moreover, the scale of the remote sensing data used is strictly linked to the reasons for the project. In other words, for some purposes, such as for global projects, coarse resolution data must be used and the reliability depends on the quality of the production process and on the consequent quality of the product itself, given the scale required for the purposes of the project.

8.2 Quality of Land-Cover Databases

In a recent book published by the Office for Official Publication of the European Communities, *Global Land Cover Validation Recommendations for Evaluation and Accuracy* (Strahler et al. 2006), the authors give important recommendations and say: "As a guideline, producing a global land cover map should consist of three more-or-less equal parts: data preparation, classification, and validation. Without proper validation, any land cover map, whether at global, regional, or local scale, remains an untested hypothesis."

Then, in another part of the same document, the authors write: "We urge map producers, as well as funding agencies, to accept the challenge of providing proper, statistically-based accuracy assessments. A validation plan and sample design should be part of every proposed and funded effort to map global land cover."

Indeed, it is often the case that only minimal resources are devoted to quality control of the photointerpretation (or classification) process and to the validation of the database, that is the assessment of the level of agreement between the database and another representation of reality which is considered more reliable.

Due to cost and time, quality control of the photointerpretation as well as validation can only be performed on the basis of a sample of polygons or points in the methodological framework of statistical inference.

When polygons are delineated by photointerpretation, quality control should be performed by repeating the production process for a sample of polygons using the same basic material and the same procedure.

In order to validate a land-cover map, a sample of polygons is compared with the corresponding ground truth, provided the scale of the remote sensing data is compatible with ground truth; otherwise, the comparison is generally made with other remote sensing data taken at a more detailed scale.

If a land-cover map is produced by the semiautomatic classification of remote sensing data, its quality can be controlled and validated using a variety of units (e.g., pixels, blocks of pixels or polygons, when delineated).

"A practical problem associated with statistical validation is the high cost of carrying out a global probabilistic sampling design, both in the effort required to collect and analyze a sufficient sample and in acquisition of the data such as ground data or the fine-resolution imagery (depending on the scale) that make it possible" (Strahler et al. 2006). Thus, sometimes other approaches are followed; for example, existing datasets are used.

Generally, when a statistical sample design is adopted for validation, it is performed in the framework of design-based inference, due to the minimal assumptions required to justify the validity of the quality estimators and their precision. This characteristic is important when many uses and users can be foreseen and assumptions must be explicit and accepted by all users.

"An inference framework heavily dependent on a model or other assumptions would require the cumbersome task of not only explicitly identifying these assumptions and model structures, but also justifying that they were satisfied for the particular application. The multitude of uses and users of a global map would suggest that validating assumptions may be even more difficult because of the large number of different analyses to which the data would be subject. Lastly, the objectivity provided by the randomization protocol of probability sampling provides assurance that the sample has not been selected, either consciously or unconsciously, to produce favorable accuracy results." (Strahler et al. 2006).

The result of the validation process is the level of accuracy, which is measured by various kinds of parameters. A common one is the total percent correct, which can be evaluated on pixels, blocks of pixels or polygons. Often, 85 percent correct (along with a variation in accuracy across the classes that is not too large) is considered an acceptable level of accuracy. However, some applications may require higher accuracy; for example, when the map is used to estimate the areas of the various land-cover types through pixel counting or the sum of the areas of the polygons (see Carfagna and Gallego 2005).

8.3 Statistical Quality Control by Acceptance Sampling

8.3.1 Classical Methods

Following the previous brief analysis of the main aspects of the quality control and validation of land-cover databases, let us now recall the main concepts of the classical theory of statistical quality control in order to show how some quality control procedures for land-cover databases can be formalized and improved by taking advantage of existing quality control methods.

When we talk about controlling the quality of a process or a manufactured product, there are a lot of aspects which are involved in the evaluation of quality, so there is no unique definition of quality. In general, a product must conform to the requirements and preferences of the consumers, and this purpose is achieved when the product respects fixed standards of production and specified levels of acceptance.

The modern approach to the definition of the quality of a product (Montgomery 2001) focuses on the concept of variability. In every production process, a certain amount of variability is present, and this implies that no unit produced is exactly equal to another: if we were to measure the same characteristic in each product, we would note a certain variability among the units. If this variability exceeds certain limits, the final product will not respect the expected characteristics and consequently will not meet the needs of the consumers. Therefore, quality improvement is achieved if there is a reduction in the variability of the process or product considered. Since variability can be expressed in statistical terms, the use of statistical techniques is essential in order to identify the causes of undesirable process behavior.

One of the oldest aspects of quality assurance is acceptance sampling, which was widely used during the 1930s and 1940s. It concerns the inspection of items from a given lot, e.g., a sample of raw materials or finished products, in order to decide whether to accept or reject the whole lot.

Several different sampling plans are used in acceptance sampling. The simplest plan is single sampling, where a random sample of units is selected from the production lot and some quality characteristics of the products are inspected. With double sampling plans, an initial sample is selected and a decision is taken to either accept the lot, reject it, or select a second sample; in the latter case, the information collected from both samples is used to decide whether to accept or refuse the lot.

Multiple sampling is a generalization of double sampling and consists of selecting, if necessary, more than two samples to make a final decision about the inspection.

This terminology is not standard, since some authors talk of double and multiple sampling while others call these two-phase and multiple-phase sampling or two-stage and multiple-stage sampling.

8.3.2 Sequential Acceptance Sampling

In quality control by acceptance sampling, the maximum number of samples is fixed in advance. In sequential acceptance sampling, a sequence of samples is selected from the lot and, at each stage, a decision is taken about whether to accept or reject the lot or whether to select a further sample. This process continues until a decision to either accept or reject the lot is made. Theoretically, the sequential sampling may continue indefinitely, until the while lot has been inspected. If the sample size at each step is equal to one, this procedure is usually called *item-by-item sequential sampling*. If the sample size at each step is greater than one, the procedure is defined as *group sequential sampling*. The item-by-item sequential sampling procedure can be illustrated by means of a Cartesian diagram where the abscissa is the total number of items selected up to that time, and the ordinate is the total number of defective items. The boundaries of acceptance and rejection are drawn on the basis of the *sequential probability ratio test* theory developed by Wald (1947). If the plotted points stay within the boundaries, another sample is selected; if a point falls above the upper line, the lot is rejected; if a point falls below the lower line, the lot is accepted.

What makes the sequential procedure different from the usual single sample inspection is the fact that the number of observations required for the sequential approach is not predetermined, since at any stage of the process the decision to terminate the inspection depends on the results of the previous observations. In general, the sequential procedure requires an expected number of observations that is considerably smaller than the fixed number of observations needed by the classical sampling procedure: Wald (1947) showed that the sequential probability ratio test leads to an average saving of about 50% in the number of observations required as compared with the most powerful classical test with the same errors of the first and second kind as the sequential test.

In the section that follows, we will examine how the quality of land-cover databases is actually controlled. We will see that very few statistical principles are applied in actual practice. In later sections, we will illustrate some proposals of ours that were inspired by both sequential acceptance sampling methods and adaptive sampling (see Thompson and Seber 1996).

8.4 Examples of Quality Control and Land-Cover Databases Validation

So far, only a few land-cover products have been validated using statistical sampling. Sometimes, stratified random sampling is adopted in order to estimate class-specific accuracy. Regions are also taken into account when the stratification is created, in order to estimate region-specific accuracies. For small or very important classes, proportional allocation of the sample is not appropriate since it does not guarantee an adequate accuracy assessment; thus, a higher sampling fraction is adopted for these classes. Budget constraints often limit the number of strata that can be effectively employed.

8.4.1 The International Geosphere–Biosphere Programme: Global Land-Cover Data Set

The International Geosphere–Biosphere Programme: Data and Information System (IGBP-DIS) DISCover (Version 1.0) 1 kilometer Global Land-Cover Data Set was submitted for validation of thematic accuracy (see Scepan 1999). Landsat Thematic Mapper and SPOT satellite data were used as a more reliable representation of reality. 379 sample units (pixels) were selected from the IGBP DISCover product using a stratified random sampling procedure. The goal was to verify a minimum of 25 pixels per DISCover class; this was accomplished for 13 of the 15 verified classes; 2 out of the 17 IGBP classes ("water and snow" and "ice") were excluded from validation. Three regional expert image interpreters independently verified each sample unit, and a majority decision rule was used to determine the accuracy. For the 15 DISCover classes validated, the average class accuracy was 59.4% with the per-class accuracy ranging from 40 to 100%. If the areas of the various classes are taken into account, an area-weighted accuracy can also be computed. The overall area-weighted accuracy of the data set was found to be 66.9 percent. This accuracy level includes geolocation errors.

8.4.2 The Global Land Cover Map 2000 Validation and Quality Control

Another example of a land-cover map validated through a probabilistic sample design is the Global Land Cover Map 2000 (GLC2000, Mayaux et al. 2006). The general objective of the European Commission's "Global Land Cover 2000" was to provide, for the year 2000, a harmonized land-cover database covering the whole globe. To achieve this objective, GLC 2000 used the VEGA 2000 data set: 14 months of preprocessed daily global data acquired by the VEGETATION instrument onboard SPOT 4.

The sampling strategy used was a two-stage stratified sampling. The stratification was based on the proportion of priority classes and on the landscape complexity. The two-stage clustering was selected due to clear advantages in terms of cost and applied on the Landsat World Reference 2 System (WRS-2). In order to assign each GLC2000 pixel to one and only one Landsat scene, whatever the latitude, Voronoi polygons (primary sampling units—PSUs) were computed from the WRS-2 centroids. PSUs were selected according to systematic sampling on an irregular stratification with different sampling rates for each stratum.

For each selected Voronoi polygon, five boxes of 3×3 km (the secondary sampling units) were selected. Boxes were chosen for the interpretation in order to reduce the impact of geolocation errors. Each 3×3 km box was interpreted according to a series of classifiers describing the basic parameters of the landscape (vegetated/unvegetated, natural/artificial, dominant layer), the water conditions (regime, seasonality, quality), and details about the tree, shrub and grass layers (cover, height, leaf type, and phenology). When the box was covered by many spatially distinct land-cover classes, the two largest classes were described along with the fraction of the box covered by each type. Then, each box was translated to the GLC2000 legend to measure the accuracy. 554 homogeneous sample sites were selected and the area-weighted global accuracy was 68.6%.

A type of quality control that does not appear to be very statistically sound, based on a qualitative comparison with ancillary data, was also performed to get an idea of the overall quality of the global product through a quick survey. The qualitative quality control was based on a systematic descriptive protocol in which each cell of the map was visually compared with reference material. The grid size was adapted to the characteristics of the landscape, the map, and the reference material. Each cell examined during the quality control procedure was characterized by a few parameters: the composition and the spatial pattern of the cell, its comparison with other existing global land-cover products, the overall quality of the cell, and the nature of any problems.

8.4.3 CORINE Land Cover

The first Corine land-cover (CLC) inventory for the EU-15 and most of the new member states was implemented between 1985 and 1996. It was carried out in order to characterize the land surface. A uniform nomenclature across Europe at a scale of 1:100,000 was used. The CLC nomenclature mostly included land-cover items, though land-use elements could also be found.

When the first update of the CLC database was performed (i.e., CLC2000), several improvements over the first inventory were introduced to improve the quality of the process. The CLC Technical Team (under the responsibility of the European Topic Centre on Terrestrial Environment) carried out a validation at the end of the project as well as a quality control during the production process (see European Environment Agency 2006).

Validation was not performed by acquiring new ground data. The LUCAS 2001/2002 survey originally carried out for agroenvironmental purposes was used instead. The accuracy of the CLC database was assessed by reinterpreting the LUCAS field photographs (in combination with IMAGE2000 and other LUCAS statistics), which were provided for 8,231 locations in the 18×18 km sampling grid. The total percent correct was 87.0 ± 0.8. However, since LUCAS was not originally intended to validate Corine Land Cover, 22 of the 44 CLC classes could not be validated due to low representativeness in LUCAS; thus, the reliability of CLC for half of the classes could not be evaluated by LUCAS. Moreover, the LUCAS survey was available only for 18 of the 29 countries where CLC was created.

The quality control during the production process was meant to monitor and provide guidelines about where to improve the production of the CLC database in the different countries. The feedback given at this stage was qualitative and its overall objective was to realize a homogeneous and comparable database at European level. The main actions supporting the production process were:

- The training of a local team of photointerpreters
- Verification of the database after 50% of the area had been produced in order to identify problems and to ensure pan-European comparability in the output
- Final analysis of the technical quality (e.g., topology, valid codes, completeness, documentation) of the different data sets to assess whether they fulfil the standards defined at the beginning of the project in order to ensure the integration of the different national databases into a common European database.

8.4.4 The ISTAT Experiment

In the projects described above, the validation procedure respects a statistical criterion, although with some difficulties; instead, the quality control during the production process is qualitative, does not utilize any statistical method, and its main aim is to check whether the database shows great differences between the various areas.

Now let us talk about an example of quality control during the production process which follows the statistical procedures of acceptance quality control (Carfagna and Gallego 1998).

In 1999, the Italian Statistical Institute (ISTAT) carried out an experiment funded by Eurostat. ISTAT produced a land-cover/land-use database with a detailed CORINE legend and a scale of 1:25,000 for the Arezzo province. The aim of the experiment was to test the difficulties that could potentially be encountered when such a detailed database is created for the whole of Italy. In this experiment, as well as qualitative analysis, the statistical acceptance quality control of lots of polygons was carried out using a parametric hypothesis test. The null hypothesis was that the lot respected the specified quality parameters. The characteristics of the polygons under scrutiny were the class of the legend assigned to the polygon, the location and the borders of the polygon. The controller, a more experienced photointerpreter, repeated the photointerpretation for a sample of polygons, and a polygon was considered incorrect if:

- The class of the legend attributed by the controller was different from the one assigned by the photointerpreter
- The difference between the areas of the two polygons (the one delineated by the controller and the one of the photointerpreter) was larger than the minimum mapping unit
- After overlaying the two polygons, the distance between the two borders was larger than 25 m
- The controller picked out, within the polygon of the photointerpreter, at least one homogeneous area that was larger than the minimum mapping unit

Due to the fact that the probability of erroneous classification of polygons is influenced by the land-cover/land-use class, as well as to guarantee the quality control of each class, a stratified sampling was adopted where the strata were the classes of the legend. Three lots were created, and in each lot 120 polygons were scrutinized. The null hypothesis was represented by the acceptable level of quality, and the alternative hypothesis was an error that was larger than or equal to 12.5%. A binomial probability distribution was assumed, the probability of erroneously rejecting the null hypothesis (the first kind of error) was fixed at 0.04, and the probability of erroneously refuting the alternative hypothesis (the second kind of error) was 0.10; thus the lot was accepted if the number of incorrect polygons was less than or equal to 10. Selected polygons were photointerpreted by the controller. All of the lots had fewer than ten incorrect polygons and were accepted (Carfagna and Napolitano 2000).

8.5 Unbiased Estimates of the Quality Parameters with Adaptive Sequential Sampling

The efficiency of sequential sample designs for acceptance control can be improved by taking advantage of the information collected during the survey through an adaptive approach.

Adaptive sampling refers to designs in which the selection of the units is based on the values of the variables of interest observed during the survey. Therefore, the sampling plan changes during the course of the survey.

In general, adaptive procedures are more complicated to design and analyze than conventional ones, and in some settings they are more difficult to implement. Moreover, adaptive sampling introduces biases into conventional estimators, so new unbiased estimators are needed.

However (Thompson 1992; Thompson and Seber 1996), adaptive sampling has a major advantage: the enhanced gathering of important observations, which can result in higher quality estimates of parameters such as the mean and the variance of the variable of interest. Therefore, for a given sample size and cost, more valuable information can be obtained than is possible with conventional designs, sometimes significantly so.

Since a sequential adaptive procedure allows a hypothesis to be tested with a sample size that is not fixed in advance and is generally smaller than in the case of classical sampling, we have proposed an adaptive sequential sampling procedure for testing whether the production process does not respect the specified quality and thus should be modified as soon as possible during the production (Carfagna and Marzialetti 2007). The aim of this approach is to monitor and improve the production process at the same time. In order to save economic resources, the same sample units used when monitoring the production process should be used to estimate the quality of the product, which should be expressed by one or more suitable parameters.

Accurate estimates of these parameters should be given. However, an adaptive sequential procedure does not produce unbiased estimates, for the same reasons that it allows very efficient hypothesis testing: data collected on already selected sample units drive the selection of successive sample units and determine the sample size, through a sample selection stopping rule linked to the parameters to be estimated.

Therefore, our aim has been to create an adaptive sequential procedure that allows us to reach decisions with the smallest sample size and in the shortest time in order to continuously improve the production process and, at the same time, produce efficient and unbiased estimates of the quality (see Carfagna 2007a, 2007b; Carfagna and Marzialetti 2007).

We have achieved this aim by adopting:

- An efficient stopping rule that is not linked to the parameter to be estimated
- An adaptive sequential method for computing the sample size in stratified sampling with Neyman's allocation (Cochran 1977) that depends on the data collected in previous steps
- A sequential selection procedure for sample units that is independent of the data collected in previous steps

We have proposed stratified sampling in order to guarantee that all of the characteristics represented by the different strata are controlled and to improve the sampling efficiency. We have adopted Neyman's allocation, which maximizes the efficiency of the sample design; moreover, the parameters that guide Neyman's allocation are continuously updated by adaptive estimates during the various steps of the sequential procedure.

In order to guarantee that the data collected in the previous steps influence the sample size but not the sample selection, in each stratum we have proposed that the sample units should selected according to the permanent random numbers method (Ohlsson 1995): "Each unit in the list frame is assigned a random number drawn independently from the uniform distribution on the interval $[0, 1]$. Let X_i denote the random number assigned to unit i. The frame units are sorted in ascending order of the X_i. The sample is composed of the first n units in the ordered list. Ohlsson (1992) presents a formal proof that this technique produces a *srswor*" (simple random sampling without replacement).

8.6 An Adaptive Sequential Procedure

We have created an adaptive sequential procedure which guarantees efficient monitoring of the production process, efficient and unbiased estimates of the quality parameters, and continuous improvement of the production process (see Carfagna and Marzialetti 2007).

Let us briefly describe the adaptive sequential procedure. A first stratified random sample is selected with a probability that is proportional to the stratum size, using the permanent random numbers method and selecting at least two sample units from each stratum. Call n the sample size for the whole area; it is small, since the main aim of this first sample is to produce first estimates for the standard errors of the variable to be estimated in the various strata in order to compute Neyman's allocation with sample size $n + 1$.

If in one stratum the estimate of the variability is zero, we do not know whether this result is due to an absence of variability in the stratum or to the low sample size; thus, the variance estimated in the stratum with the lowest positive variance is assigned to the stratum with zero variance. Otherwise, when the stratum is small and not very important, it can be merged into a similar one with positive variance.

Then, Neyman's allocation is computed with a sample size of $n + 1$. The difference between the number of sample units assigned by the two allocations is computed, and one unit is selected in the stratum where the sample size is farthest below the size assigned by Neyman's allocation.

The quality parameter and its precision are estimated. If the precision is acceptable, the process stops; otherwise, Neyman's allocation is computed with the sample of size $n + 2$, and a sample unit is selected in the stratum with the maximum difference between actual allocation and Neyman's allocation. Then the corresponding precision of the quality parameter is computed and tested, and so on, until the precision considered to be acceptable is reached. At each step of the process, the estimates of standard deviation which guide the allocation are updated.

The aim of this procedure is to select the smallest sample that allows the pre-assigned precision of the estimate to be reached. When the sample size is pre-assigned, the sequential procedure stops when this sample size is reached, although the precision can be lower then the chosen one, and the aim of the adaptive sequential procedure is then to maximize the efficiency of the sample allocation.

8.7 Two-Step Adaptive Procedure

In some cases, the parameters that guide Neyman's allocation cannot be updated continuously during the adaptive sequential procedure. Thompson and Seber (1996, Sect. 8.2.3) faced the problem of sample allocation without previous information on the variability inside the strata by suggesting a stratified random survey in two phases or, more generally, in k phases (phases are sampling steps), which allows the variability by the first (or previously selected) samples to be estimated.

The estimator of the total at the k-th phase is unbiased if at the k-th phase a complete sample is selected (each of the strata need to be sampled), with allocation based on the variability inside the strata estimated by the data collected in the previous phases. The weighted average of the estimators of the total in the various phases is an unbiased estimator if the weights are fixed in advance and do not depend on observations made during the survey and if units are selected from each stratum at each phase. These conditions have a negative effect on the efficiency of the procedure. Thus, Carfagna (2007b) proposed adaptive sampling in two steps with permanent random numbers.

A permanent random number is assigned to all sampling units in each stratum. Then, a first stratified random sample of size n_1 is selected with a probability proportional to stratum size. As in the sequential approach, the main aim of this first sample is to estimate the standard errors of the quality parameter, so its sample size is small. However, unlike the sequential procedure, the standard deviations in the various strata are estimated only once—they are not sequentially updated, so the first sample size should be large enough to produce reliable estimates of the total sample size that allow the prefixed precision to be reached and efficient sample allocation.

Once the standard deviations of the quality parameter in the different strata have been estimated, the total sample size $(n_1 + n_2)$ corresponding to the desired standard deviation can be computed, as well as Neyman's allocation. In some strata, the optimum sample size (n_h) can be less than or equal to the number of sample units already selected (n_{1h}). In such cases, due to the use of permanent random numbers, no other sample units are selected, unlike Thompson and Seber's approach, where a complete sample has to be selected at each step.

Carfagna (2007b) proved that, in the approach described here, the sample size and the sample allocation are influenced by the data collected, but the sample selection is not. Thus the two-step adaptive procedure with permanent random numbers guarantees that the direct expansion estimator is unbiased and more efficient than the two-phase estimator proposed by Thompson and Seber.

If the sample size is pre-assigned, the advantage offered by the proposed two-step procedure is the efficient allocation of the sample among the various strata.

8.8 Continuous Improvement of a Database of Remote Sensing Data

The theoretical research described above was triggered by the need to develop procedures for continuously improving databases of polygons derived from the photointerpretation of remote sensing data when quality control is centralized.

The controller, a very experienced photointerpreter, repeats the photointerpretation on selected polygons, and a polygon is considered incorrect if conditions like the ones stated in the Istat experiment occur (concerning the locations of the polygons, the distance to be considered depends on the scale of the remote sensing data).

The quality is characterized by some parameters such as the area, the percentage of the area, and the percentage of polygons correctly photointerpreted. We have sequentially estimated the parameters of interest using a stratified sample with Neyman's allocation for performing continuous quality control during the production of the database, not after its completion. The aim is to continuously improve the production process by modifying its parameters (Carfagna 2007a; Carfagna and Marzialetti 2007).

We have performed several simulations corresponding to different permanent random number selections and applied our adaptive sequential procedure. The result is that the standard deviations of the estimates decrease as the sample size increases, although this decrease is not strictly monotonous, as showed in Fig. 8.1.

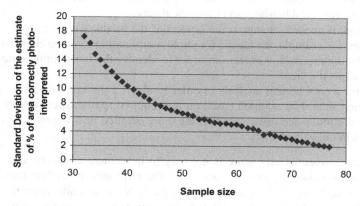

Fig. 8.1 Standard deviation of the percentage of area correctly photointerpreted for different sample sizes

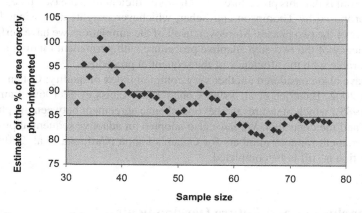

Fig. 8.2 Convergence of the estimator to the percentage of area correctly photointerpreted in the population

Figure 8.2 shows that with increasing sample size, and consequently with decreasing standard deviation, the estimator tends to converge to the value of the percentage of area correctly classified in the population, which is 83.67%.

We have compared this procedure with stratified sampling with proportional allocation and fixed sample size. We found that adaptive sequential sampling is much more efficient.

We have also proposed an analogous sequential procedure for continuous validation during the production process in order to:

- Detect discrepancies between the database and reality that make the product inappropriate for the customer's needs in a timely manner
- Evolve the characteristics of the product such that they progressively become closer to the customer's needs; note that the customer is often not aware of his/her requirements until he/she starts using the database
- Test and, if needed, change some aspects of the production process in a timely manner; for example, the legend for photointerpretation could be changed during the photointerpretation process in order to identify the most appropriate legend for the specific area
- Perform a cost–benefit analysis in order to identify the kind of remote sensing data that should be adopted based on the required spatial, spectral and temporal resolution, since using data that are more detailed than required by the user results in unjustified costs

Indeed, the sequential procedure is not easy to implement when the validation is performed by collecting ground data, since the sequential procedure requires the continuous updating of the parameters that guide Neyman's allocation in order to select the next polygon to be validated; thus, we proposed the two-step adaptive procedure with permanent random numbers described above.

The result is that this procedure is much more efficient than the two-phase sample design proposed by Thompson and Seber, whichever weights are assigned to the estimators of the two phases. Moreover, in all of the simulations we have performed, the efficiency of the two-step adaptive procedure with permanent random numbers is comparable with the efficiency of the sequential procedure.

We have also considered another very common index of quality: Cohen's kappa (Agresti 2002; Banerjee et al. 1999; Cohen 1960; Fleiss et al. 1969; Tanner and Young 1985), which measures the beyond-chance agreement between the photointerpreter and the controller. We have also adopted an adaptive sequential procedure with permanent random numbers for Cohen's kappa (Carfagna et al. 2008). This investigation is still to be completed.

8.9 Conclusions and Future Developments

In this chapter, we have shown that, although the importance of the quality control and validation of land-cover databases has been clearly stated in recent years by important authors, procedures based on randomly selected sample data acquired on purpose are rare.

In most cases, quality control and validation are based on qualitative and subjective judgment. In these cases, quantitative measurements of the quality and the accuracy of the database, which express the reliability of the quality measurement, cannot be estimated with a certain level of precision.

We have described how statically sound procedures for quality control and validation can be performed, and we have suggested efficient methods for continuously improving the production process in order to satisfy the user's needs.

Possible future developments in this research concern the effect on Cohen's kappa index of an adaptive sequential procedure with permanent random numbers, and the adaptation of the experimental design approach to the quality control and validation of land-cover databases in order to identify the main variables that affect the production process and tune the parameters of the process better.

Acknowledgements We wish to thank Prof. Alessandra Giovagnoli and Prof. Francisco Javier Gallego for their helpful comments and suggestions as well as their generous help.

References

Agresti, A.: Categorical Data Analysis. Wiley, New York (2002)

Banerjee, M., Capozzoli, M., McSweeney, L.: Beyond kappa: a review of interrater agreement measures. Can. J. Stat. **27**, 3–23 (1999)

Birkett, M.A., Day, S.J.: Internal pilot studies for estimating sample size. Stat. Med. **13**, 2455–2463 (1994)

Carfagna, E.: Innovazione continua nella elaborazione di dati telerilevati per la gestione del territorio. Proc. Intermediate Meeting of the Research Project of Relevant National Interest "Statistica e tecnologia a sostegno delle imprese". Department of Statistics, Bologna, 15–16 Feb. (2007a)

Carfagna, E.: Crop area estimates with area frames in the presence of measurement errors. Proceedings of ICAS-IV, 4th Int. Conf. on Agricultural Statistics, Advancing Statistical Integration and Analysis (paper invited by Michael A. Steiner), Beijing, 22–24 Oct. (2007b)

Carfagna, E., Gallego, J.F.: Thematic Maps and Statistics (invited paper). In: Land Cover and Land Use Information Systems for European Union Policy Needs, Office for Official Publications of the European Communities, Luxembourg, pp. 111–121, ISBN 92-828-74450-8 (1998)

Carfagna E., Gallego J.F.: Using remote sensing for agricultural statistics. Int. Stat. Rev. **73**, 3, 389–404 (2005)

Carfagna, E., Marzialetti, J.: Sequential design in quality control and validation of land cover data bases. In: Vicario, G., Isaia, E.D. (eds.): Proc. Joint ENBIS-DEINDE 2007 Conf. "Computer Experiments versus Physical Experiments," Torino, Italy, 11–13 April (2007); paper submitted to J. Appl. Stoch. Mod. Bus. Ind. (ASMBI)

Carfagna, E., Marzialetti, J., Maffei, S.: Sequential and two phase sample designs for quality control. Proc. XLIV Sci. Meeting of the Italian Statistical Society, University of Calabria, Italy, 25–27 June (2008)

Carfagna, E., Napoletano, P.: Statistica e cartografia per la creazione e l'utilizzo di basi di dati sull'uso del suolo. Proc. XL Sci. Meeting of the Italian Statistical Society, Florence, Italy, 26–28 April 2000, pp. 747–750 (2000)

Cohen, J.: A coefficient of agreement for nominal scales. Educ. Psychol. Meas. **20**, 37–46 (1960)

Cochran, W.C.: Sampling techniques, 3rd edn. Wiley, New York (1997)

European Environment Agency: The thematic accuracy of Corine land cover 2000—assessment using LUCAS. EEA Tech. Rep. 7 (2006)

Fleiss, J.L., Cohen, J., Everitt, B.S.: Large sample standard errors of kappa and weighted kappa. Psychol. Bull. **72**, 323–327 (1969)

Mayaux, P., Strahler, A., Eva, H., Herold, M., Shefali, A., Naumov, S., et al.: Validation of the Global Land Cover 2000 Map. IEEE Trans. Geosci. Rem. Sens. **44**(7), 1728–1739 (2006)

Montgomery, D.C.: Introduction to statistical quality control, 4th edn. Wiley, New York (2001)

Ohlsson, E.: Coordination of samples using permanent random numbers. In: Cox, B., Binder, D., Chinnapa, B., Christianson, A., Colledge, M., Kott P. (eds.): Business survey methods. Wiley, New York, pp. 153–169 (1995)

Ohlsson, E.: SAMU—The system for co-ordination of samples from the business register at Statistics Sweden—a methodological description (R&D Rep. 1992:18). Statistics Sweden, Stockholm (1992)

Scepan, J.: Thematic validation of high-resolution global land-cover data sets. Photogramm. Eng. Rem. S. **65**, 1051–1060 (1999)

Strahler, A.S., Boschetti, L., Foody, G.M., Friedl, M.A., Hansen, M.C., Herold, M., et al.: Global land cover validation recommendations for evaluation and accuracy assessment of Global Land Cover Maps (EUR 22156 EN). Office for Official Publication of the European Communities, Luxembourg (2006)

Tanner, M.A., Young, M.A.: Modeling agreement among raters. J. Am. Stat. Assoc. **80**, 175–180 (1985)

Thompson, S.K.: Sampling. Wiley, New York (1992)

Thompson, S.K., Seber, G.A.F.: Adaptive Sampling. Wiley, New York, ISBN 0-471-55871-0 (1996)

Wald, A.: Sequential Analysis. Wiley, New York (1947)

Chapter 9
An Innovative Online Diagnostic Tool for a Distributed Spatial Coordinate Measuring System

Fiorenzo Franceschini, Maurizio Galetto,
Domenico Maisano, and Luca Mastrogiacomo

Abstract There is currently an increasing trend for accurate measurements of large-scale lengths; in particular, 3D coordinate metrology at length scales of 5 m to 60 m has become a routine requirement in industries such as aircraft and ship construction. This chapter focuses on the Mobile Spatial coordinate Measuring System (MScMS), a new system developed at the Industrial Metrology and Quality Engineering Laboratory of DISPEA of the Politecnico di Torino. Based on a distributed sensor network structure, MScMS is designed to perform simple and rapid indoor dimensional measurements of large-size objects. Using radiofrequency (RF) and ultrasound (US) signals, the system makes it possible to localize—in terms of spatial coordinates—the points "touched" by a wireless mobile probe. To protect the system from potential causes of error, such as US signal diffraction and reflection, external uncontrolled US sources (key jingling, neon blinking, etc.) or unacceptable software solutions, MScMS implements some statistical tests in order to perform online diagnostics. One of these tests is analyzed in depth in this chapter: the "energy model-based diagnostics test." Although it is specifically developed for the MScMS

Fiorenzo Franceschini
Department of Production Systems and Business Economics (DISPEA)
Politecnico di Torino, Corso Duca degli Abruzzi 24, 10129, Torino, Italy
e-mail: fiorenzo.franceschini@polito.it

Maurizio Galetto
Department of Production Systems and Business Economics (DISPEA)
Politecnico di Torino, Corso Duca degli Abruzzi 24, 10129, Torino, Italy
e-mail: maurizio.galetto@polito.it

Domenico Maisano
Department of Production Systems and Business Economics (DISPEA)
Politecnico di Torino, Corso Duca degli Abruzzi 24, 10129, Torino, Italy
e-mail: domenico.maisano@polito.it

Luca Mastrogiacomo
Department of Production Systems and Business Economics (DISPEA)
Politecnico di Torino, Corso Duca degli Abruzzi 24, 10129, Torino, Italy
e-mail: luca.mastrogiacomo@polito.it

Pasquale Erto, *Statistics for Innovation*
ISBN 978-88-470-0814-4, © Springer 2009

system, this test can easily be extended to other recent large-scale metrology systems based on distributed devices—such as the Metris indoor-GPS, the Metronor Portable CMM, and the 3rd Tech Hi-Ball.

9.1 Introduction

In many industrial fields (automotive, aerospace, etc.), it is necessary to quickly and easily take dimensional measurements of large-size objects (Bosch 1995; Cauchick-Miguel et al. 1996; Hansen and De Chiffre 1999; Franceschini et al. 2007; Franceschini and Galetto 2007). At present, this problem can be handled using various metrological systems based on different technologies (optical, mechanical, electromagnetic, etc.). These systems are more or less adequate, depending on the measuring conditions, the user's experience and skill, the cost, accuracy, portability, etc. When measuring medium-to-large-size objects, portable systems are generally preferred to fixed ones. Transferring the measuring system to the location of the object to be measured is often more practical than moving the object to the measuring system (Bosch 1995).

The performances of most measuring systems, independent of their technology and features, can be affected by several sources of error, such as temperature, humidity, light, vibrations, etc. For this reason, the use of diagnostic tools to control measuring activities and to assist in the detection of abnormal functioning can be very helpful.

This chapter analyzes some online diagnostic tools implemented in the Mobile Spatial coordinate Measuring System (MScMS) that can be used to continuously monitor the reliability of its measurements.

MScMS, which was developed at the Industrial Metrology and Quality Engineering Laboratory of DISPEA of the Politecnico di Torino, is based on a distributed sensor network structure (Franceschini et al. 2008b). The system is designed to perform dimensional measurements of medium-to-large-size objects (longerons of railway vehicles, airplane wings, fuselages, etc.). These objects are difficult to measure using traditional coordinate measurement systems such as coordinate measurement machines (CMMs) because of their limited working volumes (ISO 10360, part 2 2001; Bosch 1995). The working principle of MScMS is very similar to that adopted by the well-known NAVSTAR GPS (NAVigation Satellite Timing And Ranging Global Positioning System) (Hofmann-Wellenhof et al. 2001). The main difference is that MScMS is based on US technology aimed at evaluating spatial distances instead of RF. MScMS is easily adaptable to different measuring environments and does not require complex procedures for installation, start-up or calibration (Franceschini et al. 2008b).

Although the diagnostic tools presented in this chapter are specifically developed for the MScMS system, they can be easily extended to other recent large-scale metrology systems consisting of distributed devices, such as the Metris indoor-GPS, the Metronor Portable CMM and the 3rd Tech Hi-Ball (Metris 2008; Metronor 2008; Welch et al. 2001).

9.2 The Concept of the "Reliability of a Measurement"

When we refer to the field of CMMs, the concept of "online metrological per-formance verification" is strictly related to the notion of "online self-diagnostics" (Gertler 1998; Franceschini and Galetto 2007). In some senses, this approach is complementary to that of uncertainty evaluation (ISO/TS 15530–6 2000; Phillips et al. 2001; Savio et al. 2002; Piratelli-Filho and Di Giacomo 2003; Feng et al. 2007). In general, an online measurement verification is a guarantee of the preser-vation of a measurement system's characteristics (including accuracy, repeatabil-ity and reproducibility) (VIM 2004; GUM 2004). The effect of measuring system degradation is the production of unreliable measurements.

In general, we can define the concept of the "reliability of a measurement" as follows.

For each measurable value x, we can define an acceptance interval $[\mathrm{LAL}, \mathrm{UAL}]$ (where LAL stands for lower acceptance limit and UAL for upper acceptance limit): $\mathrm{LAL} \leq x \leq \mathrm{UAL}$.

The measure y of the quantity x, obtained by a given measurement system, may be considered the realization of a random variable Y. It is considered "reliable" if $\mathrm{LAL} \leq y \leq \mathrm{UAL}$.

Therefore, the I and II type probability errors (misclassification rates) corre-spond, respectively, to:

$$\alpha = \Pr\{Y \notin [\mathrm{LAL}, \mathrm{UAL}] \,|\, \mathrm{LAL} \leq x \leq \mathrm{UAL}\} \tag{9.1}$$

and

$$\beta = \Pr\{\mathrm{LAL} \leq y \leq \mathrm{UAL} \,|\, x \notin [\mathrm{LAL}, \mathrm{UAL}]\} \tag{9.2}$$

from the point of view of the measurement system.

LAL and UAL are not usually known a priori.

The acceptance interval is defined by considering the metrological characteristics of the measurement system (accuracy, reproducibility, repeatability, etc.), as well as the required quality level of the measurement result (VIM 2004; GUM 2004).

The problem of online system self-diagnostics is not a recent matter, and many strategies have been proposed in different fields to address it (Clarke 1995; Henry and Clarke 1992; Isermann 1984). In the most critical sectors, such as the aeronau-tical and nuclear ones, where there is an absolute need to promptly detect every malfunction, the typical approach is based on "physical redundancy". This princi-pally consists of instrumentation and system control device replication. Although effective, this method can affect system cost and complexity (Gertler 1998).

An alternative and/or complementary method to physical redundancy is "model-based redundancy" (also called "analytical redundancy"). This approach replaces the replication of physical instrumentation with the use of appropriate mathematical models. Such models may be derived from applying physical laws to experimental data or from self-learning methods (for example, neural networks). These kinds of diagnostics allow the detection of system failures by comparing measured and model-elaborated process variables (Gertler 1998; Reznik and Solopchenko 1985; Franceschini and Galetto 2007).

The basic idea behind the self-diagnostic method described in this chapter is to define an acceptance interval. If the measurement value (y) is included in this interval, the acceptance test gives a positive response and the measured result is considered reliable. Otherwise, the measurement is rejected (Franceschini et al. 2008a).

After a general description of MScMS, the chapter focuses on the online diagnostic tool. A numerical example is presented and discussed. The following aspects are analyzed in detail: a theoretical description of the test; empirical definitions of the test parameters and acceptance limits; trial runs and preliminary experimental results; critical aspects and possible improvements.

9.3 MScMS Technological and Operating Features

The MScMS prototype is made up of three main components (see Fig. 9.1) (Franceschini et al. 2008b):

- A constellation (network) of wireless devices ("Crickets"), which are opportunely distributed around the working area
- A measuring probe that communicates via ultrasound transceivers (US) with constellation devices in order to obtain the coordinates of the touched points
- A computing and control system (PC), which receives and processes data sent by the mobile probe in order to evaluate the geometrical features of objects

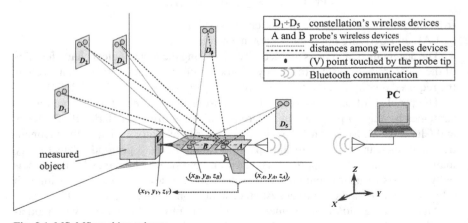

Fig. 9.1 MScMS working scheme

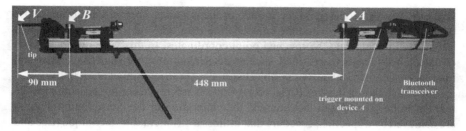

Fig. 9.2 Mobile probe prototype. The distance between the two probe devices is a construction parameter defined during the probe design phase

The measuring probe comprises a mobile system that hosts two wireless devices, a tip to touch the surface points of the measured objects, and a trigger to activate data acquisition (see Fig. 9.2) (Franceschini et al. 2008b).

Given the geometrical characteristics of the mobile probe, the tip coordinates can be univocally determined by means of the spatial coordinates of the two probe Crickets (Franceschini et al. 2008b).

The Crickets are being developed by the Massachusetts Institute of Technology and Crossbow Technology Inc. They utilize one radiofrequency (RF) and two ultrasound (US) transceivers in order to communicate and evaluate mutual distances (see Fig. 9.3) (MIT CSAIL 2004). Mutual distances are estimated by a technique known as TDoA (time difference of arrival) (Gustafsson and Gunnarsson 2003). RF communication allows each Cricket to rapidly find out distances between the devices. A Bluetooth transmitter connected to one of the two probe Crickets sends this distance information to the PC, which is equipped with ad hoc software that can analyze it.

The system makes it possible to calculate the location—in terms of spatial coordinates—of the object points that are "touched" by the probe. More precisely,

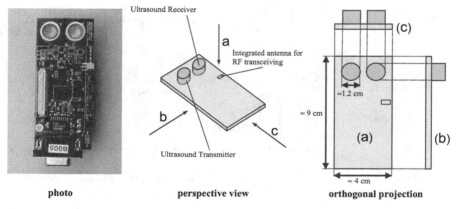

Fig. 9.3 Cricket structure (MIT CSAIL 2004); reproduced here with permission

when the trigger mounted on the mobile probe is pulled, the current distances between the probe Crickets and the constellation ones are sent to the PC. These data are utilized to calculate the touched point coordinates. In this way, different types of calculations can be performed, such as determinations of distances, geometrical tolerances, geometrical curves or object surfaces (Franceschini et al. 2008b).

The constellation devices (Crickets) operate as reference points (beacons) for the mobile probe. The spatial location and the calibration of the constellation devices are achieved by a specific procedure that utilizes a "trilateration" technique (Lee and Ferreira 2002a, 2002b; Franceschini et al. 2008b).

To uniquely determine the location of a point in 3D space, at least four reference points are generally needed (Chen et al. 2003; Sandwith and Predmore 2001; Akcan et al. 2006). In general, a trilateration problem can be formulated as follows. Given a set of N nodes with known coordinates (x_i, y_i, z_i), $i = 1, \ldots, N$, and a set of measured distances d_{M_i} from a given point $P \equiv (x_P, y_P, z_P)$, the following system of nonlinear equations needs to be solved to compute the unknown coordinates (x_P, y_P, z_P) of P (see Fig. 9.4):

$$
\begin{bmatrix}
(x_1 - x_P)^2 + (y_1 - y_P)^2 + (z_1 - z_P)^2 \\
(x_2 - x_P)^2 + (y_2 - y_P)^2 + (z_2 - z_P)^2 \\
\vdots \\
(x_N - x_P)^2 + (y_N - y_P)^2 + (z_N - z_P)^2
\end{bmatrix}
=
\begin{bmatrix}
d_{M_1}^2 \\
d_{M_2}^2 \\
\vdots \\
d_{M_N}^2
\end{bmatrix}
\tag{9.3}
$$

If this trilateration problem is over-defined (i.e., four or more reference points are available), it can be solved using a least mean squares approach (Savvides et al. 2001).

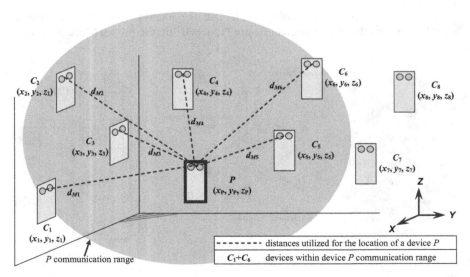

Fig. 9.4 Location of a generic device P

The location of each unknown node can be estimated by performing iterative minimization of the following error function ($\text{EF}(\mathbf{x}_P)$) (Franceschini et al. 2008b):

$$\text{EF}(\mathbf{x}_P) \equiv \frac{\sum_{i=1}^{N} (d_{C_i} - d_{M_i})^2}{N} \tag{9.4}$$

where:

- N is the number of reference points $\mathbf{x}_i = (x_i, y_i, z_i)$, $i = 1, \ldots, N$ known a priori
- $\mathbf{x}_P = (x_P, y_P, z_P)$ are the unknown coordinates of the point P in the localization space $\xi \subseteq \mathbb{R}^3$
- d_{M_i} is the measured distance between the i-th reference point and P
- d_{C_i} is the Euclidean distance between the i-th reference point and P:

$$d_{C_i} = \sqrt{(x_P - x_i)^2 + (y_P - y_i)^2 + (z_P - z_i)^2} . \tag{9.5}$$

The problem of finding the minimum of the function $\text{EF}(\mathbf{x}_P)$ can be treated as the problem of finding the point of equilibrium for a mass–spring system (lowest potential energy) (Moore et al. 2004; Franceschini et al. 2008a).

As an example, let us consider the 2D situation described in Fig. 9.5. A unitary mass is associated with each network node. The node with an unknown location is connected to three reference nodes by three springs. Each of these has a rest length equal to the measured distance and a unitary force constant.

Knowing the rest lengths (d_{M_i}) and the locations of the masses, the system potential energy is given by:

$$U(\mathbf{x}_P) = \sum_{i=1}^{N} \frac{1}{2} \left(\sqrt{(x_P - x_i)^2 + (y_P - y_i)^2} - d_{M_i} \right)^2 . \tag{9.6}$$

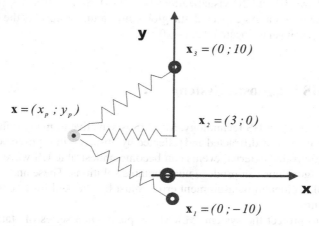

Fig. 9.5 An example of 2D mass–spring system. Three reference nodes $(\mathbf{x}_1, \mathbf{x}_2, \mathbf{x}_3)$ with known locations are linked by springs to the point to be localized $(\mathbf{x}_P)$

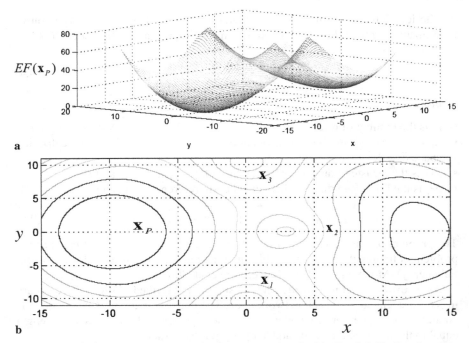

Fig. 9.6 a EF($\mathbf{x}_P$) behavior for the mass–spring system described in Fig. 9.5. Finding the minimum point means localizing the node P that has an unknown location. **b** Isoenergetic curves for the mass–spring system described in Fig. 9.5. Note that $\mathbf{x}_P$ is the global minimum point of potential energy. The maxima correspond to the reference points ($\mathbf{x}_1, \mathbf{x}_2, \mathbf{x}_3$). *Black curves* refer to low energy levels, *gray curves* refer to high energy levels

Figure 9.6 shows 3D and 2D visualizations of EF($\mathbf{x}_P$). Since EF($\mathbf{x}_P$) $\propto U(\mathbf{x}_P)$, they have the same minima. As expected, the global minimum represents the position of the node that we wish to locate ($P \equiv (-10; 0)$).

9.4 MScMS Diagnostic System

Since it is based upon US technology, MScMS is sensitive to many influential factors. US signals may be diffracted and reflected by obstacles interposed between two devices, uncontrolled external events can become undesirable US wave sources, or positioning algorithms can produce unacceptable solutions. These and other potential causes of accidental measurement errors must be checked for to ensure proper levels of accuracy.

In order to protect the system, MScMS implements a series of statistical tests for online diagnostics. The one analyzed in this chapter is the "energy model-based diagnostics test."

9.5 Energy Model-Based Diagnostics

$\text{EF}(\mathbf{x}_P)$ is non-negative by definition (see Eq. 9.6). In particular, $\text{EF}(\mathbf{x}_P) = 0$ when $d_{M_i} = d_{C_i}$, for $i = 1, \ldots, N$. Because of the natural variability of the measuring instrument, two typical situations may occur:

- $\text{EF}(\mathbf{x}_P)$ is strictly positive, even at the correct point of localization.
- $\text{EF}(\mathbf{x}_P)$ shows a global minimum at a point that is not the correct one. In other words, due to the "noise" in distance measurements, a local minimum may turn into a global minimum and vice versa.

Energy model-based diagnostics introduces a criterion in order to identify all unacceptable minima solutions for $\text{EF}(\mathbf{x}_P)$ and thus prevent system failures. Such a criterion enables the MScMS system to distinguish between reliable and unreliable measurements.

Consider a solution $\mathbf{x}_P^*$ to the problem $\min_{\mathbf{x}_P \in \xi} \text{EF}(\mathbf{x}_P)$. In general, if the problem is overdetermined (i.e., there are more than three distance constraints in the 3D case and more than two in the 2D case) and the individual measurements are affected by noise, the solution that satisfies all distance constraints at the same time does not exactly fit the real location of the node (see Fig. 9.7).

In such a case, the differences between measured and Euclidean distances may be defined as residuals ($\varepsilon_i \equiv (d_{M_i} - d_{C_i})$). Generally, in the absence of systematic sources of error, it is reasonable to hypothesize a normal distribution for the random variables ε_i, i.e.:

$$\varepsilon_i \equiv (d_{M_i} - d_{C_i}) \sim N\left(0, \sigma_i^2\right) . \tag{9.7}$$

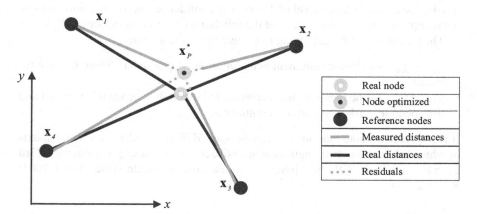

Fig. 9.7 An example of possible node localization. Measured distances are not equal to real distances

If $\sigma_i^2 = \sigma^2$, $\forall i$ (this is true in the absence of spatial/directional effects), Eq. 9.4 becomes:

$$EF(\mathbf{x}_P) = \sum_{i=1}^{N} \frac{(d_{M_i} - d_{C_i})^2}{N} = \sum_{i=1}^{N} \frac{\varepsilon_i^2}{N} = \frac{\sigma^2}{N} \cdot \sum_{i=1}^{N} \frac{\varepsilon_i^2}{\sigma^2} = \frac{\sigma^2}{N} \cdot \sum_{i=1}^{N} \left(\frac{\varepsilon_i}{\sigma}\right)^2 = \frac{\sigma^2}{N} \cdot \sum_{i=1}^{N} z_i^2.$$

(9.8)

Equation 9.8 can be seen as the sum of N independent normal squared random variables with zero mean and unit variance, multiplied by the constant term $\frac{\sigma^2}{N}$.

It should be noted that the sum in Eq. 9.8 has only $N - 1$ independent terms. Equation 9.8 causes the loss of a degree of freedom. This implies that, once $N - 1$ terms are known, the N-th one is univocally determined.

When χ_P^2 is defined as $\chi_P^2 = \sum_{i=1}^{N} \left(\frac{\varepsilon_i}{\sigma}\right)^2$, $EF(\mathbf{x}_P)$ in Eq. 9.8 has a chi-square distribution with $N - 1$ degrees of freedom:

$$EF(\mathbf{x}_P) = \frac{\sigma^2}{N} \cdot \chi_P^2.$$

(9.9)

The residual variance σ^2 can be estimated a priori for the whole measuring space, for example during the phase of installation and calibration of the system.

Every time a measurement is performed for each probe Cricket, the MScMS diagnostic software computes the following quantity (experimental chi-square):

$$\chi_P^{2*} = EF(\mathbf{x}_P^*) \frac{N}{\sigma^2}.$$

(9.10)

Assuming that the risk α is a type I error, a one-sided confidence interval for variable $\chi_{v,\alpha}^2$ can be calculated. $\chi_{v,\alpha}^2$ is a chi-square distribution with $v = N - 1$ degrees of freedom and a confidence level of $1 - \alpha$. The confidence interval is assumed to be the acceptance interval for the test of the reliability of the measurement.

The test arrives at the following two alternative conclusions:

- $\chi_P^{2*} \leq \chi_{v,\alpha}^2 \rightarrow$ the measurement is not considered unreliable; hence it is not rejected
- $\chi_P^{2*} > \chi_{v,\alpha}^2 \rightarrow$ the measurement is considered unreliable; hence it is rejected and the operator is asked to perform another one

It is important to note that this test can be applied in many other different contexts in which trilateration or triangulation are utilized for coordinate measurement (3rd Tech Hi-Ball, Leica T-Probe, Metris Laser Radar and i-GPS, etc.) (Welch et al. 2001; Rooks 2004).

9.5.1 Setting Up the Test Parameters

The risk α is defined by the user according to the required level of performance of the system. A high value of α prevents unacceptable solutions to the optimization problem, minimizing the type II error β.

On the other hand, while a low value of α speeds up the measurement procedure, it may result in inaccurate data being collected due to the high level of II type error β.

The estimation of the residual variance can be evaluated in two ways: by applying the uncertainty composition law to the calculation of the coordinates, starting from the measurement uncertainty of the distances between the constellation beacons and the probe crickets (GUM 2004), or empirically, on the basis of experimental distance measurements. In this case, it is estimated from a sample of residuals obtained by measuring a set of points that are randomly distributed across the whole working volume. This method requires knowledge of the locations of the measured points a priori. It can be easily implemented during the initial phase of setting up and calibrating the system.

In the following, we focus on this second estimation procedure.

Given a set of M points distributed in the measurement space $\xi \subseteq \mathbb{R}^3$, randomly measured by a single Cricket (i.e., with a random sequence of measurements and a random position and orientation of the Cricket), a set of N_j residuals can be calculated for each point j, $j = 1, \ldots, M$.

It should be noted that the number of residuals N_j may change due to the different number of distances detected during each measurement.

In the absence of systematic sources of error and time or spatial/directional effects, it is reasonable to hypothesize the same normal distribution for all the random variables ε_{ij}, $j = 1, \ldots, M$, $i = 1, \ldots, N_j$, i.e.:

$$\varepsilon_{ij} \equiv (d_{M_i} - d_{C_i})_j \sim N(0, \sigma^2) . \tag{9.11}$$

The variance σ^2 may be estimated as follows:

$$\hat{\sigma}^2 = \sum_{j=1}^{M} \sum_{i=1}^{N_j} \frac{(\varepsilon_{ij} - 0)^2}{\sum_{j=1}^{M} N_j} = \sum_{j=1}^{M} \sum_{i=1}^{N_j} \frac{(\varepsilon_{ij})^2}{\sum_{j=1}^{M} N_j} . \tag{9.12}$$

The value obtained for $\hat{\sigma}^2$ is considered the reference value for the test.

With this notation, Eq. 9.10 becomes:

$$\chi_P^{2*} = \text{EF}(\mathbf{x}_P^*) \cdot \frac{N}{\sigma^2} \cong \text{EF}(\mathbf{x}_P^*) \cdot \frac{N}{\hat{\sigma}^2} . \tag{9.13}$$

9.5.2 An Example of the Application of Energy Model-Based Diagnostics

A preliminary empirical investigation was carried out to verify the accuracy of this approach.

Considering that ultrasound sensors are able to achieve uncertainties of about 10 mm for distance measurements (confidence level $1 - \alpha = 0.95$, i.e., a covering factor $k \cong 2$, according to GUM 2004) in a network consisting of five reference points (constellation beacons) placed in the measurement volume schematized in Fig. 9.8, $\hat{\sigma}^2$ was empirically estimated as follows:

- $M = 253$ points randomly distributed in the working volume were measured by a single Cricket.
- The coordinates $\mathbf{x}_j$, $j = 1, \ldots, M$, of each node were evaluated using the "mass–spring" localization algorithm and a sample of 1123 residuals were obtained.
- A normal residual distribution was tested using a chi-square test (Montgomery 2005).
- The residual variance was estimated by Eq. 9.12. The value obtained was $\hat{\sigma}^2 = 100.0 \, \text{mm}^2$ (see Table 9.1 for a summary of the data).

The acceptance limit for $\mathrm{EF}(\mathbf{x}_P)$, assuming a type I risk $\alpha = 0.05$ and $v = N - 1 = 5 - 1 = 4$ degrees of freedom, is:

$$\mathrm{EF}(\mathbf{x}_P^*) \leq \frac{\hat{\sigma}^2}{N} \cdot \chi^2_{v=4, \alpha=0.05} \Rightarrow \mathrm{EF}(\mathbf{x}_P^*) \leq 189 \, \text{mm}^2 \tag{9.14}$$

Consider now a typical situation that can occur when the ultrasound technology is used to estimate distances: US reflection. Referring to the configuration in Fig. 9.9, suppose that a generic point P inside the measurement volume (for example, $P \equiv (1067.2; -122.5; 925.8)$) has to be localized.

Table 9.1 Details of data analysis for estimating the standard deviation of the residuals

Sample size: $N_{\mathrm{TOT}} = \sum\limits_{j=1}^{M} N_j$	1123
Estimate for the mean: $\hat{\mu} = \sum\limits_{j=1}^{M} \sum\limits_{i=1}^{N_j} \dfrac{\varepsilon_{ij}}{\sum\limits_{j=1}^{M} N_j}$	0.3 mm
Estimate for the variance: $\hat{\sigma}^2 = \sum\limits_{j=1}^{M} \sum\limits_{i=1}^{N_j} \dfrac{(\varepsilon_{ij})^2}{\sum\limits_{j=1}^{M} N_j}$	100.0 mm^2
Maximum: $\varepsilon_{\mathrm{MAX}} = \max \left\{ \varepsilon_{ij} \mid i = 1, \ldots, N_j, j = 1, \ldots, M \right\}$	42.7 mm
Minimum: $\varepsilon_{\mathrm{MIN}} = \min \left\{ \varepsilon_{ij} \mid i = 1, \ldots, N_j, j = 1, \ldots, M \right\}$	−37.1 mm

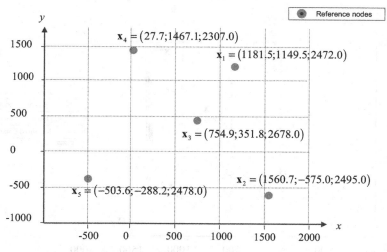

Fig. 9.8 Scheme showing the positions of the reference nodes (constellation beacons) in the measurement volume (point coordinates in millimeters [mm])

A Cricket positioned in P is able to correctly measure distances from all of the reference nodes except for one of them. An obstacle (for example, the operator performing the measurement) is interposed between P and that node, preventing direct US signal propagation. At the same time, a wall placed close to the two nodes causes US signal reflection. The consequence is that the estimate for the pairwise distance between those two nodes is 100 mm larger.

The measured distances are:

$$d_{M_1} = 2104.8 \, \text{mm}$$
$$d_{M_2} = 1713.4 \, \text{mm}$$
$$d_{M_3} = 1831.4 \, \text{mm}$$
$$d_{M_4} = 2355.6 \, \text{mm}$$
$$d_{M_5} = 2215.2 \, \text{mm} \tag{9.15}$$

In this case, the algorithm produces the following wrong localization solution (see Fig. 9.10): $\mathbf{x}_{P'}^* \equiv (1022.6; -187.3; 911.8)$, characterized by a high level of "energy:" $\text{EF}(\mathbf{x}_P^*) \cong 904 \, \text{mm}^2 > 189 \, \text{mm}^2$.

Because of this result, the energy model-based diagnostics indicate that the measurement should be rejected.

Upon removing the obstacle, the distance from beacon 1 becomes $d_{M_1} = 2004.8 \, \text{mm}$, and we obtain the correct localization solution:

$$\mathbf{x}_P^* \equiv (1067.2; -122.5; 925.8) . \tag{9.16}$$

The new "energy" value is: $\text{EF}(\mathbf{x}_P^*) \cong 41 \, \text{mm}^2 < 189 \, \text{mm}^2$, so $\mathbf{x}_P^*$ cannot be considered unreliable and the measurement is not rejected.

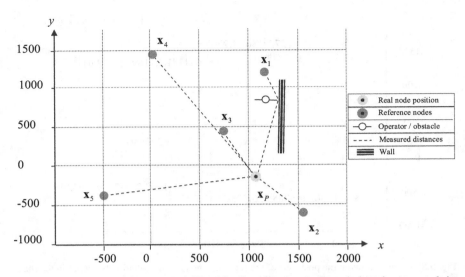

Fig. 9.9 Scheme illustrating a potentially misleading situation: walls and obstacles can result in wrong distance estimates (point coordinates are given in millimetres [mm]; see Fig. 9.8). In this case, the measured distance between node 1 and node P is higher than the actual distance

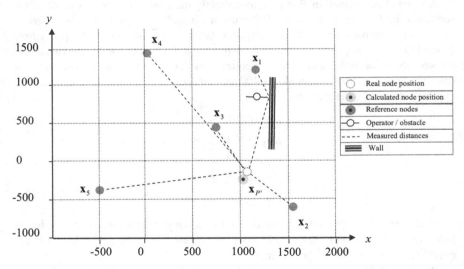

Fig. 9.10 Scheme illustrating a wrong localization solution (P') due to a wrong estimate for the distance between node 1 and node P (point coordinates are given in millimetres [mm]; see Fig. 9.8)

9.6 Conclusions

MScMS is an innovative wireless measuring system that is complementary to CMMs. A prototype of this system has been developed at the Industrial Metrology and Quality Engineering Laboratory of DISPEA of the Politecnico di Torino. It is portable, not very expensive, and suitable for large-scale metrology (which is not easy to perform with conventional CMMs).

Some innovative aspects of the system concern its online diagnostic tools. When dealing with measurement systems, good measurement diagnostics are crucial to applications in which errors can lead to serious consequences.

The diagnostics tool described in this chapter, which is based on the concept of the "reliability of a measurement," enables MScMS users to reject measurements which do not satisfy a statistical acceptance test with a given confidence coefficient.

After rejection, the operator is asked to perform the measurement again, changing the orientation/positioning of the probe; or, if necessary, to rearrange the beacons in the system network.

Preliminary results from the application of this online diagnostic tool reveal that it exhibits acceptable efficiency in preserving the system from measurement failures. However, in some cases the system requires the measurement to be repeated too many times, resulting in excessive duration of the measuring process.

Future work, as well as improvements in the power of the existing tools, will be aimed at enriching the MScMS control system by implementing additional tools that are able to steer the operator during measurement. For example, they could suggest the position of the probe in the measuring volume, or propose possible extensions to the network of beacons, or automatically filter and/or correct corrupted measurements.

References

Akcan, H., Kriakov, V., Brönnimann H., Delis A.: GPS-free node localization in mobile wireless sensor networks. In: Proceedings of MobiDE'06, Chicago, IL, USA, 25 June (2006)

Bosch, J.A.: Coordinate Measuring Machines and Systems. Marcel Dekker, New York (1995)

Cauchick-Miguel, P., King, T., Davis, J.: CMM verification: a survey. Measurement 17(1), 1–16 (1996)

Chen, M., Cheng, F., Gudavalli, R.: Precision and accuracy in an indoor localization system (Technical Report CS294-1/2). University of California, Berkeley, CA (2003)

Feng, C.X.J., Saal, A.L., Salsbury, J.G., Ness, A.R., Lin, G.C.S.: Design and analysis of experiments in CMM measurement uncertainty study. Precis. Eng. 31(2), 94–101 (2007)

Franceschini, F., Galetto, M.: A taxonomy of model-based redundancy methods for CMM on-line performance verification. Int. J. Technol. Manage. 37(1–2), 104–124 (2007)

Franceschini, F., Galetto, M., Maisano, D.: Management by Measurement—Designing Key Indicators and Performance Measurement Systems. Springer-Verlag, Berlin (2007)

Franceschini, F., Galetto, M., Maisano, D., Mastrogiacomo, L.: A review of localization algorithms for distributed wireless sensor networks in manufacturing. Int. J. Computer Integr. Manuf. (in press) (2008a). doi:10.1080/09511920601182217

Franceschini, F., Galetto, M., Maisano, D., Mastrogiacomo, L.: Mobile Spatial coordinate Measuring System (MScMS)—introduction to the system. Int. J. Prod. Res. (in press) (2008b). doi:10.1080/00207540701881852

Gertler, J.J.: Fault Detection and Diagnosis in Engineering Systems. Marcel Dekker, New York (1998)

GUM: Guide to the Expression of Uncertainty in Measurement. International Organization for Standardization, Geneva (2004)

Gustafsson, F., Gunnarsson, F.: Positioning using time difference of arrival measurements. Proc. IEEE Int. Conf. on Acoustics, Speech, and Signal Processing (ICASSP 2003), Hong Kong 6, 553–556 (2003)

Hansen, H.N., De Chiffre, L.: An industrial comparison of coordinate measuring machines in Scandinavia with focus on uncertainty statements. Precis. Eng. 23(3), 185–195 (1999)

Hofmann-Wellenhof, B., Lichtenegger, H., Collins, J.: GPS. Theory and Practice. Springer, Wien (2001)

ISO: 10360, part 2: Geometrical Product Specifications (GPS)—acceptance and reverification tests for coordinate measuring machines (CMM). International Organization for Standardization, Geneva (2001)

ISO/TS: 15530-6 (Working Draft): Geometrical product specifications (GPS)—coordinate measuring machines (CMM): techniques for determining the uncertainty of measurements. Part 6: Uncertainty assessment using uncalibrated workpieces. International Organization for Standardization, Geneva (2000)

Lee, M.C., Ferreira, P.M.: Auto-triangulation and auto-trilateration. Part 1. Fundamentals. Precis. Eng. 26(3), 237–249 (2002a)

Lee M.C., Ferreira P.M.: Auto-triangulation and auto-trilateration—Part 2: Three-dimensional experimental verification. Precis. Eng. 26(3), 250–262 (2002b)

Metris: Webpage. http://www.metris.com/large_volume_tracking__positioning/ (2008)

Metronor: Webpage. http://www.metronor.com (2008)

MIT Computer Science and Artificial Intelligence Lab: Cricket v2 User Manual. http://cricket.csail.mit.edu/v2man.html (2004)

Montgomery, D.C.: Introduction to Statistical Process Control. Wiley, New York (2005)

Moore D., Leonard J., Rus D., Teller S.S.: Robust distributed network localization with noisy range measurements. Proceedings of SenSys 2004, Baltimore, MD, pp. 50–61, 3–5 Nov. (2004)

Phillips, S.D., Sawyer, D., Borchardt, B., Ward, D., Beutel, D.E.: A novel artifact for testing large coordinate measuring machines. Precis. Eng. 25(1), 29–34 (2001)

Piratelli-Filho, A., Di Giacomo, B.: CMM uncertainty analysis with factorial design. Prec. Eng. 27(3), 283–288 (2003)

Reznik, L.K., Solopchenko, G.N.: Use of a-priori information on functional relations between measured quantities for improving accuracy of measurement. Measurement 3(3), 98–106 (1985)

Rooks, B.: A vision of the future at TEAM. Sensor Rev. 24(2), 137–143 (2004)

Sandwith, S., Predmore, R.: Real-time 5-micron uncertainty with laser tracking interferometer systems using weighted trilateration. Proc. 2001 Boeing Large Scale Metrology Seminar, St. Louis, MO, 13–14 Feb. (2001)

Savio, E., Hansen, H.N., De Chiffre, L.: Approaches to the calibration of freeform artefacts on coordinate measuring machines. Ann. CIRP 51/1, San Sebastian, Spain, pp. 433–436 (2002)

Savvides, A., Han, C., Strivastava, M.B.: Dynamic fine-grained localization in ad hoc networks of sensors. Proc. ACM/IEEE 7th Annu. Int. Conf. on Mobile Computing and Networking (MobiCom'01), pp. 166–179, July (2001)

VIM: International Vocabulary of Basic and General Terms in Metrology. International Organization for Standardization, Geneva (2004)

Welch, G., Bishop, G., Vicci, L., Brumback, S., Keller, K.: High-performance wide-area optical tracking. The HiBall Tracking System. Presence Teleoper. Virtual Env. 10(1), 1–21 (2001)

Chapter 10
Technological Process Innovation via Engineering and Statistical Knowledge Integration

Biagio Palumbo, Gaetano De Chiara, and Roberto Marrone

Abstract This chapter shows the strategic role that a systematic approach to planning for a designed industrial experiment plays in technological process innovation. Guidelines already proposed in the literature emphasizing the pre-experimental planning phase are customized and applied in a case study concerning the laser drilling process of a combustion chamber in aerospace industry. The team approach is the real driving force for pre-experimental activities; it enables the integration of engineering and statistical knowledge, catalyzes process innovation and, moreover, it allows a virtuous cycle of sequential learning to be put into action. The innovative technological results obtained in the first screening experimental phase are presented. Since these results arise from a sound systematic approach, they enable a future experimental phase on optimization and robustness to be planned. The case study of a laser drilling process provides a best-practice guide to synergic collaboration and partnership between academic statisticians and industrial practitioners; it was developed by AVIO, an aerospace company at the leading edge of propulsion technology.

Biagio Palumbo
Department of Aerospace Engineering, University of Naples Federico II, Naples, Italy
e-mail: biagio.palumbo@unina.it

Gaetano De Chiara
AVIO S.p.A., Manufacturing Technologies Department, Pomigliano, Naples, Italy
e-mail: gaetano.dechiara@aviogroup.com

Roberto Marrone
AVIO S.p.A., Manufacturing Technologies Department, Pomigliano, Naples, Italy
e-mail: roberto.marrone@aviogroup.com

Pasquale Erto, *Statistics for Innovation*
ISBN 978-88-470-0814-4, © Springer 2009

10.1 Introduction

In industry today, there is a general awareness that one-factor-at-a-time (OFAT) experiments are always less useful than statistically designed experiments (Wu and Hamada 2000; Montgomery 2001; Box et al. 2005). Ilzarbe et al. (2008) pointed out that, although the design of experiments (DOE) methodology has been applied in industry for many years, it is still not used as it should be. In fact, many engineering applications are of little educational value, since they include the experimental matrix and the analysis of the results associated with them but lack any details about the pre-experimental steps performed.

The first practical and systematic framework that effectively tries to answer "how" to plan activities in the pre-experimental phase (i.e., the phase that precedes the actual experiment) was proposed in Coleman and Montgomery (2003), where predesign master guide sheets and supplementary sheets were designed and applied in a case study involving the CNC machining of a jet engine impeller. Quoting Coleman and Montgomery (2003, p. 2): "The guide sheets are designed to be discussed and filled out by multidisciplinary experimentation team [...];" "The sheets are intended to encourage the discussion and resolution of generic technical issue needed before the experimental design is developed;" "The guide sheets [...] outline a systematic script for the verbal interaction among the people on the experimentation team. When the guide sheets are completed, the team should be well equipped to proceed with the task of designing the experiment."

The aim of this chapter is to apply the approach proposed in Coleman and Montgomery (2003) in the context of laser drilling, highlighting the strategic role that this systematic approach plays in technological process innovation and, moreover, as the critical starting point for a virtuous cycle of sequential learning.

The process of sequential learning (Box 2001)—in other words, continuous improvement based on the Shewhart–Deming cycle (Plan-Do-Check-Act) (Deming 1982)—comes from team work in which statistical and technological competencies are fully exploited. "A new environment calls for a new strategy for statisticians" (Hahn 2007, p. 646); in order to be a "proactive statistician" (Hahn 2007) or, in other words, a "statistical colleague" (Hunter 1981), a "statistical catalyst" (Box 2001), or a "statistical leader" (Deming 1982), it is necessary to work as a team member by directly sharing in the excitement of problem solving and the responsibility for project success.

This is a chapter that involves both statistical and technological aspects. The statistical methodologies applied in this first screening experimental phase—two-level fractional factorial design and ANalysis Of VAriance (ANOVA)—are extensively treated in the literature (for example in Montgomery 2005), and so they will be applied without any explicit introduction or analytical formulation. Technological aspects related to the laser-drilling process are only briefly introduced so as to support the technological interpretation of statistical results.

The case study involving the laser-drilling process was developed by AVIO, an aerospace company at the leading edge of propulsion technology.

10.2 Technological Context and Case Study

In the aerospace industry, laser drilling is the most economical process for drilling many thousands of high-quality, small-diameter effusion holes in order to improve the cooling capacities of engine components such as blades or combustion chambers.

Three different laser-drilling methods are usually used to drill a combustion chamber: trepanning, percussion and drilling on the fly (DOF). The first of these involves cutting the circumference of the hole. The percussion method makes the hole by shooting the place to drill several times, without any relative motion between the laser and the workpiece. In the DOF method, the laser pulses are delivered to the workpiece while it is rotating around its own axis; the rotation of the part is synchronized with the laser pulse, ensuring that multiple pulses are always delivered at the exact position of the hole, so that the hole is created after a fixed number of workpiece rotations.

The DOF method is better than the others in terms of productivity (i.e., it takes less time), but it is not always better in terms of the quality of the hole. High productivity and high-quality holes are the key competitive factors for industries involved in laser-drilling processes.

The quality of the hole is related to several geometrical and metallurgical parameters (see Fig. 10.1): (a) taper (i.e., the acylindricity of the hole) and barreling (i.e., a measure of the presence of irregular depressions on the side wall of the hole); (b) recast layer (i.e., the degree of accumulation of material on the side wall of the

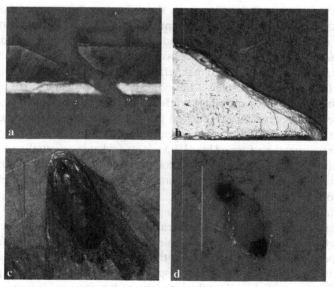

Fig. 10.1a–d Main metallurgical and geometrical defects associated with laser-drilled holes: **a** taper and barreling; **b** recast layer; **c** spatter; **d** dross

hole); (c) spatter (i.e., the amount of resolidified material at the entrance to the hole); (d) dross (i.e., the amount of resolidified material at the exit of the hole).

The case study described here concerns the drilling of a combustion chamber that has been partially coated by an internal thermal barrier. AVIO usually make the holes in the area without the thermal barrier by the DOF method; the holes in the coated area are made by trepanning because of the unsatisfactory quality of the holes achieved by the DOF method. The aim of the Manufacturing Technologies Department is to extend the DOF method to the whole combustion chamber in order to increase productivity and ensure that geometrical and metallurgical requirements continue to be met.

10.3 Pre-experimental Planning

Following the systematic approach to planning a designed industrial experiment proposed in Coleman and Montgomery (2003), two pre-design sheets (i.e., the main and secondary sheets) were conceived and implemented. We customized the proposed guide sheets in order to make them more appropriate and comprehensive in the specific technological and organizational context in which they are used. These two kinds of sheets force the experimenter to address fundamental questions from the early phases of the experimental activity and, moreover, they enable the results of the interaction between statistical and technological competences to be recorded during face-to-face discussion.

These sheets are the only official document that circulates within the team involved in the experimentation. The main sheets contain information about the objective of the experimentation, the relevant background, the response variables and the factors (i.e., control, held-constant and nuisance factors). The secondary sheets detail the technological relationship between the control factors and the response variables, in terms of the expected main effects and interactions.

In this first experimental phase, the objective was to characterize the drilling process; that is, to detect which factors affect the quality of the hole.

The relevant background was derived from previous laser-drilling experiments carried out by AVIO, bibliographic research, expert opinion, and physical laws. Previous experiments performed by AVIO adopted what is essentially an OFAT approach; unfortunately, when only one factor is varied at a time while all of the others are kept fixed, the experimental results cannot account for interactions between factors (Czitrom 1999).

Achieving optimum hole quality during laser drilling is one of most important issues in this specific research area. Several parametric studies that studied the relationship between laser parameters and hole quality characteristics for different several aerospace materials are available in the scientific literature. In particular, two papers (Low et al. 1999, 2001) offered specific contributions to the experimental activities. These papers highlight the influence of temporal pulse train modulation during laser percussion drilling of aerospace materials in terms of good overall hole geometry (i.e., taper and barreling) and metallurgical characteristics (i.e., spatter and recast layer). The new proposed temporal pulse train modulation, called "sequential pulse delivery pattern control" (SPDPC), increases the laser pulse energy linearly throughout the pulse train. This is a novel approach in laser percussion drilling; in the traditional approach, called the "normal delivery pattern" (NDP), the energy is constant in each pulse. Note that, before this, SPDPC had never been used by AVIO.

Taper and recast layer thickness were the response variables that were taken into consideration. For each variable, the normal operating level, the range, the measurement precision and the relationship to the objective were specified on the main sheets.

The study of the factors involved in the experimentation is a crucial task and requires intensive knowledge transfer. The first brainstorming step involved listing all of the factors that, according to different technological points of view and competencies, came out during team discussions. The second step consisted of classifying each factor as a control, held-constant or nuisance factor (Coleman and Montgomery 2003). Obviously, different classifications are possible, each strictly related to the specific aims of the experiment.

In this first screening experimental phase, the following control factors were adopted: peak power (A), defocus (B), pulse width (C), delivery pattern (D) and assist gas pressure (E). Factors A and C are quantitative laser pulse parameters; factor B is the distance between the laser focal spot and the workpiece surface; factor E is the pressure of the gas used in tandem with the laser beam to enhance the removal of material; factor D, as previously mentioned, is a qualitative factor that refers to the specific form of temporal pulse train modulation adopted. On the secondary sheets, for each quantitative control factor, the normal level and range as well as the measurement precision were specified. Moreover, particular attention was paid to the task of attempting to elicit the effects of each control factor as well as the effects of two-factor interactions on the response variables. Such efforts are very important when the results require technological interpretation (Sect. 10.5.2).

Held-constant factors are controllable factors whose effects are not of interest in this experimental phase. In particular, laser-drilling experiments were performed on a 250 W Nd:YAG laser emitting at a wavelength of 1.063 μm with a fixed beam delivery. The laser beam was approximately 14 mm in diameter; it was focused with a 200 mm focal length lens, giving a spot size of approximately 0.47 mm diameter (M^2 value of ∼25). Oxygen assist gas was used since the results of previous experimentation by AVIO indicated that it was the most suitable gas for this process. Each experiment was performed through a conical copper nozzle with 1.15 mm-diameter orifice, and the beam beam was always inclined at 30° to the surface, as this configuration was identified by AVIO as being the one that is most critical to this process. The materials used to produce the combustion chamber (i.e., external surface: superalloy; internal surface: thermal coating barrier) and their thicknesses are omitted here for industrial confidentiality reasons.

Nuisance factors are uncontrolled factors—factors which cannot be controlled from a technological or economical point of view. There are some potential nuisance factors that could influence the laser-drilling experiments. Laser light deterioration and the metallurgical/geometrical characteristics of the thermal barrier appear to be the most influential nuisance factors. In this first screening experimental phase, both of these factors were avoided by adopting a new laser light source and an appropriate experimental set-up (Sect. 10.4). Since the first encouraging technological results obtained (Sect. 10.5) arose from the application of a sound systematic approach, they enabled future experimental phases focused on optimization and robustness to be planned. Future experimentation will be required in order to elucidate technological and statistical aspects related to different sources of variation, which are usually classified into three categories: those external to the process, process nonuniformity, and process drift (Phadke 1989).

10.4 Experimental Design and Set-Up

In the pre-experimental phase, five factors were considered important and a 2^{5-1} design was adopted. This design, with $I = ABCDE$ (defining relation), is a resolution V design, so no main effect or two-factor interaction is aliased with other main effects or two-factor interactions, but each main effect is aliased with a four-factor interaction, and each two-factor interaction is aliased with a three-factor interaction. Since three-factor (and higher) interactions are negligible, the experimental 2^{5-1} design enables reliable information to be obtained about main effects and two-factor interactions. Table 10.1 summarizes the levels of control factors and their settings.

The delivery pattern (D) is a qualitative factor (set as NDP or SPDPC); it is related to the shape of the overall laser pulse pattern applied during the pulse train required to drill a hole. To drill a single hole by DOF, four pulses are required, one for each revolution of the workpiece. In NDP, each laser pulse has the same peak power (A) and pulse width (C). In SPDPC, each pulse may have a different peak power (A) and/or pulse width (C). In this first experimental phase, a simple laser pulse pattern (four pulses) was used; its shape is omitted here for industrial confidentiality reasons.

Table 10.2 shows the 2^{5-1} design matrix. For each treatment, three replications were executed, giving a total of 48 experimental runs. This unusual number of replications (usually two replications of each treatment are executed in a screening exper-

Table 10.1 Control factors and their settings

Control factors	Labels	Low (−)	High (+)	Unit
Peak power	A	13.5	16.7	kW
Defocus	B	0	2.5	mm
Pulse width	C	0.9	1.2	ms
Delivery Pattern	D	NDP	SPDPC	–
Assist gas pressure	E	4.4r	6.8r	bar

Table 10.2 Matrix for the 2^{5-1} design (defining relation $I = ABCDE$)

Treatment	A	B	C	D	E = ABCD
1	−	−	−	−	+
2	+	−	−	−	−
3	−	+	−	−	−
4	+	+	−	−	+
5	−	−	+	−	−
6	+	−	+	−	+
7	−	+	+	−	+
8	+	+	+	−	−
9	−	−	−	+	−
10	+	−	−	+	+
11	−	+	−	+	+
12	+	+	−	+	−
13	−	−	+	+	+
14	+	−	+	+	−
15	−	+	+	+	−
16	+	+	+	+	+

iment) was adopted to provide more consistent response repeatability, in particular for the recast layer thickness, during this first experimental phase.

The quality control test for laser-drilled microholes, in terms of taper and recast layer thickness, is a destructive test. Therefore, all of the experiments were carried out on a cylindrical artefact made out of the same material as the combustion chamber (i.e., external surface: superalloy; internal surface: thermal coating barrier). Figure 10.2 shows the application of the laser-drilling process to the: (a) combustion chamber; (b) cylindrical artefact.

In order to reduce the laser downtime and avoid technological difficulties related to laser calibration optics, only the treatment order was randomized. For each treatment, three set of holes were sequentially created at different positions on the cylindrical artefact in order to reduce the effects of lurking variables (some of which are potentially due to metallurgical and geometrical characteristics of the thermal bar-

Fig. 10.2a,b Laser drilling process in action: **a** combustion chamber; **b** cylindrical artefact

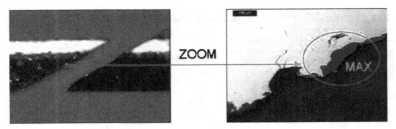

Fig. 10.3 Measurement of recast layer

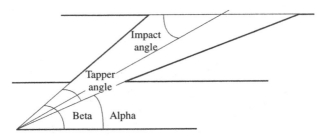

Fig. 10.4 Measurement of taper (beta – alpha)

rier). After drilling, each set of holes was cut out using the laser. For each set, one hole was chosen at random, isolated and sectioned, and then cold-mounted in resin under vacuum and polished in order to allow microscopic analysis. Finally, each obtained sample was chemically treated in order to analyze the recast layer thickness and heat-affected zone microstructures. Figure 10.1a,b shows typical images obtained by this method.

Both the taper and the recast layer thickness of the hole were measured using a microscope with different levels of zoom. The maximum recast layer thickness was measured along the longitudinal edge of the hole; see Fig. 10.3.

Taper was instead measured as the difference between the "beta" angle and the "alpha" angle, as illustrated in Fig. 10.4.

10.5 Analysis of Results and Technological Interpretation

10.5.1 Analysis of Results

The ANOVA method was applied in order to test the statistical significance of the main effects and the two-factor interactions for the taper and the recast layer. Diagnostic checking was successfully performed via graphical analysis of residuals. The experimental results for the taper and the recast layer thickness are shown, using Pareto charts of standardized effects ($\alpha = 0.05$), in Fig. 10.5.

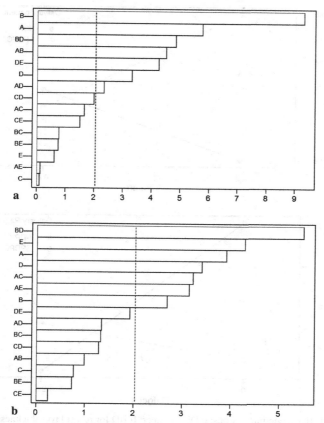

Fig. 10.5a,b Pareto charts of standardized effects ($\alpha = 0.05$): **a** taper; **b** recast layer

The two-factor interactions DE (delivery pattern–pressure) and BD (defocus–delivery pattern), for taper and recast layer, respectively, were the true "discoveries" of the first screening experimental phase. Both interactions were of the antisynergic type (Fig. 10.6) and involved the factor D (delivery pattern), up to then set to NDP by AVIO. The practitioner, even one considered expert in laser drilling, would have never been able to anticipate this type of interaction during the pre-experimental phase.

Figures 10.7 and 10.8 show the main effects plots for taper and recast layer thickness, respectively; for each response variable, the significant main effects ($\alpha = 0.05$) result from Fig. 10.5. An exhaustive technological interpretation of the results is given in Sect. 10.5.2.

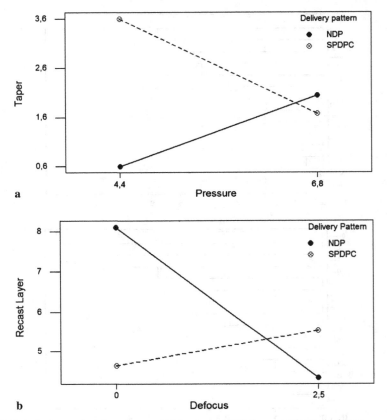

Fig. 10.6a,b Antisynergic interactions: **a** DE for taper; **b** BD for recast layer thickness

10.5.2 Technological Interpretation of Results

The technological interpretation of the results is a very important phase. Comparing the technological "expectations", elicited in the pre-experimental phase (secondary sheets) with the statistical results allows practitioners to gain technological knowledge and to determine the added value of a systematic approach to planning a designed industrial experiment. In fact, positive results from such a comparison will strengthen the trust of practitioners in experimental activities; negative results from the comparison will force the practitioner to look deeper at the technological interpretation of the results.

In Sects. 10.5.2.1 and 10.5.2.2, we consider the technological interpretation of the results for taper and recast layer thickness, respectively, in more depth.

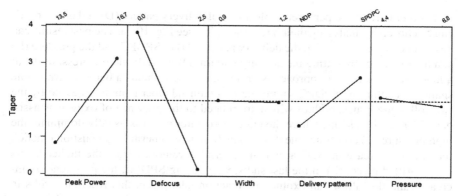

Fig. 10.7 Main effects plot for taper

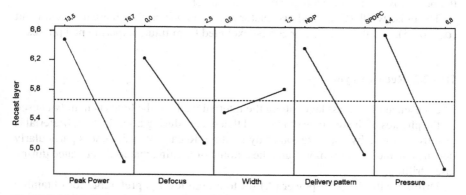

Fig. 10.8 Main effects plot for recast layer thickness

10.5.2.1 Taper

In laser drilling, the hole taper is often related to two main concurrent physical phenomena associated with the erosion of material: crater-generator and beam-divergence effects. The former occurs at the higher layers of the drilled workpiece, while the latter typically occurs at the lower layers, where the laser beam usually loses its power to both penetrate and ablate due to its natural divergence.

A low setting for the peak power (A) is preferable in order to reduce taper; see Fig. 10.7. Low peak power (A), often connected with low pulse energy, reduces the crater-generator effect as it decreases hole-wall erosion at the higher layers of the drilled piece.

A high setting for defocus (B) is able to reduce taper, Fig. 10.7. For the levels of pressure (E) and pulse energy tested, in terms of their effects on taper, the beam-divergence effect dominates over the crater-generator effect when the defocus (B) is set to a high level, while the crater-generator effect dominates when the defocus (B) is set at low level.

In order to reduce taper, preferable levels of delivery pattern (D) and pressure (E) can be chosen by analyzing their interaction plot; see Fig. 10.6a. The best results can be obtained by either setting the delivery pattern (D) to SPDPC and the pressure (E) at a high level, or by setting the delivery pattern (D) to NDP and the pressure (E) to a low level. The former approach is preferable since it yields a better internal hole shape for drilling by SPDPC. In this case, a reduced taper is mainly caused by the two following issues, both of which are related to the presence of oxygen as assist gas: (i) in the first shots, where the crater–generator effect is usually dominant, the high pressure (E) constrains the development of an exothermic combustion reaction that would increase the rate at which material is removed from the higher layers of the drilled piece; (ii) in the last shots, when using SPDPC, the pulse energy increases and this allows an exothermic reaction take place; this reaction results in more material being removed from the lower layers of the drilled piece and opposes the beam-divergence effect that would cause taper.

The pulse width (C) and all two-factor interactions involving it are not significant ($\alpha = 0.05$), and thus this factor can be excluded from future experimental phases.

10.5.2.2 Recast Layer

The presence of a recast layer along the internal walls of the holes is mainly caused by the physics of the removal of material that occurs during laser drilling. Material is usually removed by vaporization or by melting. Recast layer thickness particularly depends on the rate of removal and the amount of melting material generated during laser drilling.

Drilling with the peak power (A) set to a high level is preferable for obtaining smaller recast layers, Fig. 10.8. The heat surge of material resulting from the sudden temperature load caused by the high-energy laser pulse that occurs with high peak power (A) causes the vaporization of most of the removed material, and thus a smaller amount of recast material on the walls of the drilled holes.

Recast layer thickness can be reduced by using high levels of both defocus (B) and pressure (E). In fact, a higher level of defocus (B) promotes material removal by vaporization rather than melting, and thus enables the growth of the recast layer to be suppressed. In the one hand, high pressure (E) constrains the development of an exothermic combustion reaction that could result in the removal of more material (when less material is removed, there is less remelt material and so the thickness of the recast layer is reduced). On the other hand, a high pressure (E) increases the rate at which material is removed. In this way, the ejection of material from the drilled hole speeds up and the remelting along the internal walls of the hole decreases.

In order to reduce recast layer thickness, preferable levels of delivery pattern (D) and defocus (B) can be chosen by analyzing their interaction; see Fig. 10.6b. The best results can be obtained by either setting the delivery pattern (D) to SPDPC and the defocus (B) to a low level, or by setting the delivery pattern (D) to NDP and the defocus (B) to a high level. The former option is preferable because it reduces spatter and dross in the holes drilled by SPDPC. In this configuration, the smallest spot size and the highest energy density are obtained at the surface of the material with a low level of defocus (B). The high density promotes the removal of material by vaporization rather than melting, and this occurs mainly in the higher layers of the drilled part. On the other hand, the use of SPDPC reduces downward melt material, due to the gradual heating of the drilled part prior to any significant material removal. Both effects significantly help to reduce the recast layer thickness.

The pulse width (C) and all two-factor interactions involving it do not have a significant effect ($\alpha = 0.05$) on recast layer thickness.

10.6 Conclusion

This chapter has shown the strategic role that a systematic approach to planning for a designed industrial experiment plays in technological process innovation.

The team approach is the real driving force of pre-experimental activities; it enables engineering and statistical knowledge to be integrated, it catalyzes process innovation, and, moreover, it allows a virtuous cycle of sequential learning to be implemented.

The results obtained in this first screening experimental phase enabled the factors that affect the quality of the holes made by drilling process, in terms of taper and recast layer thickness, to be discerned. In particular, a new type of delivery pattern, that had not previously been applied by AVIO, has been successfully adopted; the antisynergic two-factor interactions between this new factor (delivery pattern) and pressure (which affects the taper)/defocus (which affects the recast layer thickness), are the true "discoveries" of this first experimental phase.

Moreover, technological interpretation of the results has allowed practitioners to gain technological knowledge and to see the added value of a systematic approach to planning for a designed industrial experiment.

Since the results obtained arise from a sound systematic approach, they enable future experimental work focused on optimization and robustness to be planned.

Acknowledgements We are grateful to other team members that contributed meaningfully to the experimental activities. In particular: S. Cinque (Manufacturing Technologies Department, AVIO) for his support in writing part of the program for laser machining; L. Gabriele (Manufacturing Technologies Department, AVIO), for his logistic, technological and motivational support; V. Genovese (Special Process Engineering Department, AVIO), for his detailed technological support in the pre-experimental phase, as well as in the technological interpretation of results.

References

Box, G.: Statistics for discovery. J. Appl. Stat. **28**(3–4), 285–299 (2001)
Box., G.E.P., Hunter, J.S., Hunter, W.G.: Statistics for Experimenters. Wiley, New York (2005)

Coleman, D.E., Montgomery, D.C.: A systematic approach to planning for a designed industrial experiment. Technometrics. **35**(1), 1–12 (1993)

Czitrom, V.: One-factor-at-a-time versus designed experiments. Am. Stat. **53**(2), 126–131 (1999)

Deming, W.E.: Out of Crisis. Cambridge University Press, Cambridge (1982)

Hahn, G.J.: The business and industrial statistician: past, present and future. Qual. Reliab. Eng. Int. **23**(6), 643–650 (2008)

Hunter, W.G.: The practice of statistics: the real world is an idea whose time has come. Am. Stat. **35**(2), 72–76 (1981)

Ilzarbe, L., Álvarez, M.J., Viles, E., Tanco, M.: Practical applications of design of experiments in the field of engineering: a bibliographical review. Qual. Reliab. Eng. Int. **24**(4), 417–428 (2008)

Low, D.K.Y., Li, L., Corfe, A.G., Byrd, P.J.: Taper control during laser percussion drilling of NIMONIC alloy using sequential pulse delivery pattern control (SPDPC). In: Proc. ICALEO: Laser Materials Processing Conf., San Diego, CA, USA, C11–C20, 15–18 Nov. (1999)

Low, D.K.Y., Li, L., Byrd, P.J.: The influence of temporal pulse train modulation during laser percussion drilling. Opt. Laser. Eng. **35**(3), 149–164 (2001)

Montgomery, D.C.: Design and Analysis of Experiments. Wiley, New York (2001)

Phadke, M.S.: Quality engineering using robust design. Prentice-Hall, London (1989)

Wu, C.F., Hamada, M.: Experiments. Wiley, New York (2000)

Part III
Innovation of Lifecycle Management

Chapter 11
Bayesian Reliability Inference on Innovated Automotive Components

Maurizio Guida and Gianpaolo Pulcini

Abstract The need to assess the reliability of new automotive products in a timely manner compels manufacturers to make use of early failure warranty data. However, the narrow observation period and the moderate sizes of early warranty data sets result in reliability estimates that are not very accurate. Nevertheless, when a new product is not revolutionary but instead the result of making improvements to its predecessors, past failure data in conjunction with corporate technical knowledge are a valuable source of information which can be usefully exploited in a Bayesian estimation framework. To this end, a Bayesian procedure was developed which is based on a rigorous formalization of both objective information provided by observed failure data for past products as well subjective information on the effectiveness of design or process modifications introduced into new products to improve their reliability. Information on modified working conditions is also formalized, and the effect of requested cost reductions for outsourced components is considered. The proposed procedure is then applied to a case study relating to a newly revised component that is assembled in a car model that was already on the market, and its ability to support reliability estimates and management decisions is addressed.

11.1 Introduction

Automotive manufacturers perform reliability analyses on their new products in order to satisfy the reliability expectations of their customers and to control the costs

Maurizio Guida
Department of Information Engineering and Electrical Engineering, University of Salerno
84084 Fisciano (SA), Italy
e-mail: mguida@unisa.it

Gianpaolo Pulcini
Istituto Motori, National Research Council, 80125 Napoli, Italy
e-mail: g.pulcini@im.cnr.it

Pasquale Erto, *Statistics for Innovation*
ISBN 978-88-470-0814-4, © Springer 2009

of repairs and substitutions during the warranty period. Reliability analyses aim to both estimate the reliability of the product and verify whether a given reliability target is attained for specified operating time t_0 (e.g., the warranty period). If the target is not achieved, then the main failure modes must be identified and design and/or process modifications that are capable of removing the causes of failure must be incorporated in a timely manner to improve the reliability of the component.

Of course, in order to minimize the undesirable consequences of a low reliability level, the reliability of the component should be accurately estimated as soon as possible.

Data sources used for reliability estimation often consist of failure data based on field observations during the warranty period ("field failure warranty data") that provide the number and the type of repairs carried out at authorized dealerships. However, when a timely estimate of product reliability is required, the available field data consist only of early warranty claims related to vehicles produced during the first few production months and observed over operating periods of less than t_0. Thus, narrow observation periods and moderate sample sizes generally produce reliability estimates that are not very accurate.

Nevertheless, many new automotive products are evolutionary, not revolutionary, since they are the result of making design improvements to their predecessors (past products). In such cases, information on past products is generally available, and this information, when properly formalized, can be used in conjunction with field data on new products in order to obtain more accurate reliability estimates.

Here, a Bayesian procedure is proposed which uses all the available information to achieve this goal for evolutionary automotive components. This procedure is based on a rigorous formalization of both the objective information provided by observed failure data on past products, and the subjective information on the effectiveness of design/process modifications introduced into new products to improve their reliability (Guida and Pulcini 2002, 2006; Guida et al. 2008). In particular, the effectiveness of each modification is measured by taking into account factors such as:

- The designer's uncertainty about the effectiveness of each planned improvement
- The environment and operating conditions of the new product, which may differ from those of the past product
- Possible requested reductions in the costs of outside suppliers

The reliability characteristics of new products at the end of the warranty period t_0 are evaluated via:

- A point and interval estimate of the reliability at t_0
- A prediction of the number of failures that will be observed over the whole warranty period in a future population of vehicles

Such reliability estimates are carried out in the following two phases:

- Before starting mass production (i.e., on the basis of past data and prior information only), in order to provide the management with a tool for deciding whether planned modifications are satisfactory or not

- During the early commercialization phase of the product, by combining prior information with field failure data observed early on in the warranty period

Finally, a case study relating to a newly revised component that is assembled in a car model that was already on the market is analyzed to illustrate the ability of the proposed procedure to support reliability estimation and decision making. The application refers to some real data that have been slightly modified for the sake of industrial confidentiality.

11.2 Prior Inference on the Failure Probability

11.2.1 Past Data

Information on the failure probability of the past product is generally provided by the fraction of failures $\bar{p}_0$ observed in the population of past products during the warranty period t_0. By assigning each component failure to one single part i ($i = 1, 2, \ldots$) of that component, past data consist of the fraction of failures $\bar{p}_{0,i}$ experienced by each part i, and the failure probability of the whole component is given by $\bar{p}_0 = \sum_i \bar{p}_{0,i}$.

Note that this can be considered the worst case, where it is assumed that no information on the vehicle age t or the number of kilometers x covered by the vehicle up to the failure of the past component is available.

11.2.2 Formalizing Modification Effectiveness

During the development phase of a new product, an attempt is made to remove the critical failure modes by making design or process modifications to its parts. When effective, such modifications will cause a reduction in the failure probability $p_{0,i}$ of the part i in the new product.

We assume that the designer is able to elicitate, for each part i, an expected limit $L_{0,i}$ for the failure probability of that part; such a limit is achieved only when the planned modification is fully effective and no new defect is introduced by this modification. This predicted value refers to the same environment and operating conditions in which the past product was working.

When the modification of part i is not fully effective, the failure probability of that part will lie within the interval bounded by the predicted limit $L_{0,i}$ and the past failure probability $\bar{p}_{0,i}$. We then assume that the designer is also able to express his "subjective" uncertainty about the ability to actually attain the planned improvement by assigning a value to the "uncertainty factor" ρ_i ($\rho_i = 1, 2, 3, 4$) on the basis of the following criteria:

1. Safe prediction ($\rho_i = 1$): the planned improvement is based on layout modifications, in which case the predicted limit will almost certainly be achieved
2. Highly likely prediction ($\rho_i = 2$): the planned modification is fairly complex, but in similar cases it produced the expected improvement; the predicted limit can then be achieved
3. Uncertain prediction ($\rho_i = 3$): the modification implies an innovative design with the support of theoretical and/or experimental studies, such as failure mode and effects analysis, fault tree analysis, ...
4. Risky prediction ($\rho_i = 4$): the modification implies an innovative design that lacks the support of theoretical and/or experimental studies

The failure probability $p_{0,i}$ of each part is then treated as a random variable that is uniformly distributed in the interval $(L_{0,i}, U_{0,i})$, where the upper limit $U_{0,i}$ depends on the observed failure probability $\bar{p}_{0,i}$ of the past product, on the predicted lower limit $L_{0,i}$, and on the "uncertainty factor" ρ_i as follows:

$$U_{0,i} = L_{0,i} + \rho_i \cdot (\bar{p}_{0,i} - L_{0,i})/4 . \tag{11.1}$$

If no modification is planned for a given part i, then the failure probability $p_{0,i}$ of such a part is assumed to be a constant equal to $\bar{p}_{0,i}$. As depicted in Fig. 11.1, for given $\bar{p}_{0,i}$ and $L_{0,i}$ values, the more risky the prediction, the higher the uncertainty about the failure probability. The failure probability p_0 of the whole new component is the sum of the $p_{0,i}$ values: $p_0 = \sum_i p_{0,i}$.

Because no further information on a possible dependence among the $p_{0,i}$ values can be formulated by the analyst, the hypothesis that the $p_{0,i}$ values are independent random variables is assumed. Under this maximum entropy hypothesis, the mean and the standard deviation of p_0 are given, respectively, by:

$$E[p_0 | \mathbf{pd}] = \sum_i (U_{0,i} + L_{0,i})/2$$

$$\sigma[p_0 | \mathbf{pd}] = \sqrt{\sum_i (U_{0,i} - L_{0,i})^2/12} , \tag{11.2}$$

where $\mathbf{pd}$ denotes the dependence of these moments on the failure data of the past product (past data).

11.2.3 Effect of Working Conditions and Cost Reduction

In Sect. 11.2.2, the expert is assumed to be able to predict the failure probability limits for the parts of the new component under the same environment and operating conditions, i.e., under the same working conditions that characterized the past product.

Fig. 11.1 Distribution of the failure probability $p_{0,i}$ of part i in the new product according to the prediction uncertainty

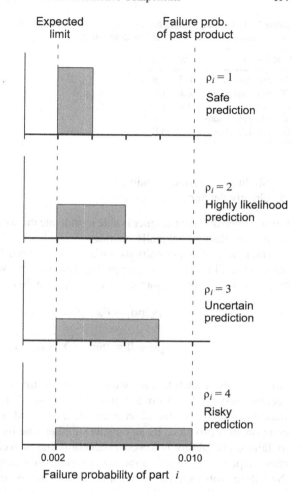

When the working conditions change, the failure probability generally changes, and so the prediction of the failure probability for the new component needs to be modified to take these changes into account.

The proposed approach accounts for the effect of modifications separately from the effect of working conditions. This strategy was chosen because the same component will generally be mounted on a number of car models that are each used in different conditions, thus implying different working conditions for the same component. By using arguments in Grahowski et al. (1976) and Vianello (1981), the working conditions are classified into five different classes:

1. Greatly improved conditions
2. Slightly improved conditions
3. Unchanged conditions

Table 11.1 Working factors for the mean and the standard deviation
of the failure probability of the new component

Conditions	δ_W	η_W
Greatly improved	0.5	0.8
Slightly improved	0.8	1.0
Unchanged	1.0	1.0
Slightly deleterious	1.2	2.0
Greatly deleterious	1.8	3.0

4. Slightly deleterious conditions
5. Greatly deleterious conditions

It is assumed that the designer is able to indicate the working conditions under which
the new product will actually operate.

The new working conditions are assumed to modify the mean and the standard
deviation in (11.2) of the failure probability p_0 of the whole component through the
multiplicative "working factors" δ_W and η_W, so that:

$$E[p_0|\mathbf{pd}] = \delta_W \cdot \sum_i (U_{0,i} + L_{0,i})/2$$

$$\sigma[p_0|\mathbf{pd}] = \eta_W \cdot \sqrt{\sum_i (U_{0,i} - L_{0,i})^2/12}, \qquad (11.3)$$

Guide values of such factors, which are the result of a large amount of experience
accumulated by Fiat Auto, are given in Table 11.1. Of course, accurate values for
such factors strictly depend on component type and on corporate experience. If the
component is produced by an outside supplier, one factor that can greatly influence
its failure probability is the cost reduction that is eventually requested of the supplier. Experience suggests that any cost reduction inevitably implies an increase in
the failure probability, even when the contractual requirements for product reliability remain unchanged. This is often because new failure modes arise that can also
involve parts that were previously free of failures.

We assume that both the predicted limit $L_{0,i}$ and the "working factors" δ_W and
η_W refer to unreduced costs (i.e., the designer quantifies the effectiveness of the
design/process modifications and the effects of different working conditions while
being unaware of the cost reduction), and that the effect of possible cost reductions
on the failure probability of the whole new component is formalized later, according
to the "size" of the requested reduction, classified as:

- A moderate cost reduction, e.g., $\leq 25\%$
- A large cost reduction, e.g., $> 25\%$

Cost reduction then acts on the failure probability p_0 by modifying the mean and
the standard deviation (11.3) through the multiplicative "cost factors" δ_C and η_C,

Table 11.2 Cost factors for the mean and the standard deviation
of the failure probability of the new component

Cost reduction	δ_C	η_C
Moderate reduction	1.5	1.5
Large reduction	2.0	2.0

respectively. Thus:

$$E[p_0|\mathbf{pd}] = \delta_C \cdot \delta_W \cdot \sum_i (U_{0,i} + L_{0,i})/2$$

$$\sigma[p_0|\mathbf{pd}] = \eta_C \cdot \eta_W \cdot \sqrt{\sum_i (U_{0,i} - L_{0,i})^2/12} \,, \tag{11.4}$$

are the predicted mean and standard deviation of the failure probability of the new
product, taking into account planned design or process modifications, environmental
and operating conditions, and possible cost reductions. Table 11.2 gives plausible
values of "cost factors." However, if the cost reduction is totally or partially due to
the removal of some parts in the new product, then the actual cost reduction must
be computed by deducting the cost of the removed parts from the cost of the whole
product before using Table 11.2.

11.2.4 Prior Inference on the Failure Probability
of the New Product

For mathematical tractability, prior information on the failure probability p_0 of the
new product, given the warranty data for the past product and all subjective information on modification effectiveness, working conditions and possible cost reductions,
is formalized through the beta density:

$$g(p_0|\mathbf{pd}) = \frac{p_0^{a-1}(1-p_0)^{b-1}}{B(a,b)}, \quad 0 \leq p_0 \leq 1 \,. \tag{11.5}$$

The mean and standard deviation of this equal the predicted values in (11.4), so that
the beta parameters a and b are:

$$a = \frac{E^2[p_0|\mathbf{pd}](1 - E[p_0|\mathbf{pd}])}{\sigma^2[p_0|\mathbf{pd}]} - E[p_0|\mathbf{pd}]$$

$$b = a\left(\frac{1}{E[p_0|\mathbf{pd}]} - 1\right) \,. \tag{11.6}$$

Figure 11.2 compares the exact distribution of $p_0 = \sum_i p_{0,i}$ (obtained by using the
values in Table 11.3 and assuming no change in working conditions and no cost
reduction) to the beta density (11.5) with a mean and a variance that match the exact
values given by (11.2). The beta approximation appears to work well.

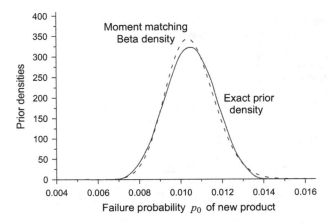

Fig. 11.2 Exact (——) and moment-matching beta (– – –) (prior) densities of the failure probability p_0 of a new product with no change in working conditions and no cost reduction

Table 11.3 Past product failure frequency and prior information on design/process modifications in the case study

Part i	$\bar{p}_{0,i}$	$L_{0,i}$	ρ_i	$U_{0,i}$
XX12A	0.00237	0	4	0.00237
XX12B	0.00096	0.00096		0.00096
XY11A	0.00048	0	1	0.00012
XY11B	0.00237	0	4	0.00237
ZY09A	0.00048	0.00048		0.00048
ZZ22A	0.00285	0	3	0.00214
ZZ22B	0.00048	0	1	0.00012
ZZ22C	0.00144	0.00144		0.00144
ZZ22D	0.00426	0.00300	2	0.00363
XZ02A	0.00048	0	4	0.00048
XZ01B	0.00048	0.00048		0.00048

In Fig. 11.3, some prior densities (11.5) that correspond to different designer's judgments about the working conditions of the new product are depicted, assuming no cost reductions and the following (arbitrary but plausible) values for the mean and the standard deviation, respectively:

$$\sum_i (U_{0,i} + L_{0,i})/2 = 0.01 \quad \text{and} \quad \sqrt{\sum_i (U_{0,i} - L_{0,i})^2/12} = 0.002 . \tag{11.7}$$

The prior density (11.5) provides a predictive estimate of the failure probability of the new product at the end of the warranty period t_0. In particular, a γ upper credibility limit on p_0, say p_γ, can be obtained by numerically solving:

$$\int_0^{p_\gamma} g(p_0|\mathbf{pd}) \, dp_0 = IB(p_\gamma; a, b) = \gamma \tag{11.8}$$

where $IB(p_\gamma; a, b)$ is the incomplete beta function.

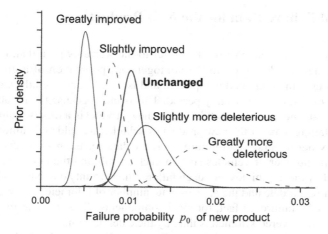

Fig. 11.3 Prior densities of the failure probability p_0, based on the judgements of different designers about the working conditions of the new product

11.2.5 Prior Prediction of the Number of Failed Items

On the basis of the prior density (11.5), the number $m(t_0)$ of items in the future new-product population that will fail during the warranty period can easily be predicted. Let N be the planned size of the future population. The predictive likelihood of $m(t_0)$, given p_0, is then:

$$P[m(t_0) = m|p_0] = \binom{N}{m} p_0^m (1 - p_0)^{N-m}, \quad m = 0, 1, 2, \ldots, \tag{11.9}$$

and the prior predictive distribution of $m(t_0)$, given past data and technical information, is obtained by integrating the product of the predicted likelihood (11.9) and the prior density (11.5) over p_0:

$$P[m(t_0) = m|\mathbf{pd}] = \int_0^1 P[m(t_0) = m|p_0] g(p_0|\mathbf{pd}) \, dp_0$$

$$= \binom{N}{m} \frac{B(a+m, b+N-m)}{B(a, b)} \tag{11.10}$$

where the dependence on past data is implicitly conveyed by the beta parameters a and b. Prior prediction limits on $m(t_0)$ can also be obtained from (11.10). For example, a conservative γ upper limit for, say $m_\gamma(t_0|\mathbf{pd})$, is the smallest value of m_U that satisfies the inequality:

$$\sum_{m=0}^{m_U} P[m(t_0) = m|\mathbf{pd}] \geq \gamma. \tag{11.11}$$

11.3 Field Failure Data for the New Product

While failure data for the past product refer to the warranty period t_0, field failure data for the new product refer to the homogeneous population of a given type of vehicle that uses the upgraded component and is mostly observed during the early part t (with $t < t_0$) of the warranty period. This occurs because the reliability team aims to estimate the failure probability of the new product in a timely manner, and so it analyzes failure data for the new product as vehicles are sold and authorized dealerships carry out repairs. Field failure warranty data for the new product typically consist of the number of vehicles sold in each month and the component failures experienced by these vehicles. For each failure, the warranty database registers both the calendar repair date and the approximate number of kilometers x covered by the vehicle up to the failure. In fact, for most components, "life" is measured better by the number x of covered kilometers than by the calendar time t.

Failure data in a warranty database are generally grouped into bands of distance traveled ("mileage") with a convenient band width (e.g., 3000 km), because the recorded mileage of the vehicle up to the occurrence of component failure is often inaccurate. In fact, when the failure is not catastrophic, the vehicle is often brought to the authorized dealership some time after the failure has occurred. In addition, a sound analysis of the warranty data requires that the number of vehicles with components that have not failed should be distributed in the mileage bands, on the basis of both the operating time t of each vehicle and ad hoc surveys on vehicle mileage distribution (Campean et al. 2001; Lu 1998; Rai and Singh 2005).

Because a vehicle is a repairable system, the warranty data may include multiple claims for the same vehicle relating to a given component. However, multiple failures usually represent a very small fraction of the failures observed in the warranty period; hence, we assume that no multiple failures are present, and that the effect of such an assumption on the analysis is negligible.

Warranty data for the new product are classified in terms of:

- The number m_k of failures in the kth mileage band (x_{k-1}, x_k) $(k = 1, 2, \ldots, l)$
- The number c_k of vehicles suspended in the kth mileage band, i.e., the number of vehicles in the population for which the component has not failed and that have covered a number of kilometers within the band (x_{k-1}, x_k) during their operating time

Hence, by assuming that the mileage covered by suspended vehicles coincides with the midpoint of the band, say $y_k = (x_{k-1} + x_k)/2$ (see, for example, Attardi et al. 2005), the likelihood function for the new data, say **nd**, is given by:

$$L(\mathbf{nd}|\mathbf{R}) \propto \prod_{k=1}^{l} \{[R(x_{k-1}) - R(x_k)]^{m_k} R(y_k)^{c_k}\} R(x_l)^{c_{l+1}} \qquad (11.12)$$

where $\mathbf{R} = (R(x_1), \ldots, R(x_l), R(y_1), \ldots, R(y_l))$ is the vector of the reliability of the new product at ages $x_1, \ldots, x_l, y_1, \ldots, y_l$, $x_0 = 0$, and c_{l+1} denotes the number of vehicles surviving beyond the last mileage band.

Before using these data to estimate the failure probability p_0 of the new product up to t_0, we must consider two issues.

First, the failure probability p_0 is not exactly the component unreliability, because the component life is measured by the mileage covered and items installed in different vehicles accumulate different mileages up to t_0. More correctly, the failure probability p_0 can be viewed as an "average unreliability" in t_0 (Lu and Rudy 2000) when averaging the unreliability function $F(x)$ with respect to the distribution $f(x|t_0)$ of the kilometers covered in t_0 over the population of the owners:

$$p_0 = \int_0^{\infty} F(x) f(x|t_0) \, dx . \tag{11.13}$$

However, the average unreliability is well approximated by the unreliability evaluated at the average mileage $\bar{x}_0 \equiv E[x|t_0]$ covered by the vehicles during t_0 (Lu 1998):

$$p_0 \simeq F(\bar{x}_0) \equiv F_0 . \tag{11.14}$$

The second consideration is that a functional form must be chosen for the reliability $R(x)$, because the new product data refer to different mileages. For a high-quality production process, infant mortality is negligible and this ensures that deterioration failures constitute most of the failures observed over the whole warranty period. In this case, the two-parameter Weibull function:

$$R(x) = \exp[-(x/\alpha)^{\beta}], \quad \alpha > 0; \, \beta > 1 \tag{11.15}$$

is a suitable choice (Guida and Pulcini 2002). From (11.14) and (11.15), we note that $\alpha^{\beta} = -\bar{x}_0^{\beta}/\ln(1 - F_0)$. Then, substituting $R(z) = (1 - F_0)^{(z/\bar{x}_0)^{\beta}}$ into (11.12), the likelihood function is reparameterized in terms of β and $F_0 \equiv F(\bar{x}_0)$:

$$L(\mathbf{nd}|F_0, \beta) \propto \prod_{k=1}^{l} [(1 - F_0)^{(x_{j-1}/\bar{x}_0)^{\beta}} - (1 - F_0)^{(x_j/\bar{x}_0)^{\beta}}]^{m_k} (1 - F_0)^{X(\beta)} , \tag{11.16}$$

where:

$$X(\beta) = \sum_{k=1}^{l} c_k (y_k/\bar{x}_0)^{\beta} + c_{l+1} (x_l/\bar{x}_0)^{\beta} .$$

The two-parameter Weibull assumption can be checked by performing a graphical analysis. A nonparametric estimate of the reliability of the new product at the upper bound of each mileage band is given by the actuarial recursive formula (Nelson 1982):

$$\tilde{R}(x_k) = \tilde{R}(x_{k-1}) \left[1 - \frac{m_k}{r_k - 0.5 c_k} \right] \tag{11.17}$$

where the term $-0.5 c_k$ adjusts for the censored units (which cover about half of the mileage band), $\tilde{R}(x_0) \equiv 1$, and $r_k = \sum_{i=k}^{l} (m_i + c_i) + c_{l+1}$ (with $k = 1, \ldots, l$) is the number of vehicles at risk at the beginning of band k. The estimated reliability values can then be plotted on Weibull paper against the upper bound x_k of each mileage band. If the model is adequate, the plot should be reasonably close to a straight line.

11.4 Posterior Inference on the New Product

11.4.1 Posterior Inference on the Reliability

The likelihood function (11.16) of the new data is now parameterized in terms of the unreliability F_0 and the shape parameter β, and hence a joint prior density $g(F_0, \beta | \mathbf{pd})$ for F_0 and β, given the data on past products, must be formalized. Under the assumption that F_0 and β are prior independent random variables, the density $g(F_0, \beta | \mathbf{pd})$ is the product of the prior density for F_0 and the prior density for β. Within the approximation (11.14), the prior density (11.5) for p_0 can be immediately converted into the prior information for F_0, such that:

$$g(F_0 | \mathbf{pd}) = \frac{F_0^{a-1}(1 - F_0)^{b-1}}{B(a,b)}, \quad 0 \leq F_0 \leq 1. \tag{11.18}$$

When formulating the "subjective" prior density for the Weibull parameter β, technical knowledge about the failure-causing mechanism can be usefully exploited. In fact, the assumption of the Weibull model relies on deterioration being the dominant failure mechanism, and this implies $\beta > 1$. Moreover, when a Weibull analysis based on past data is available, the estimated β value can be used as a guide to setting at least an interval (β_1, β_2) of plausible values for β. With no further information available, uniform density over this interval appears to be a plausible choice:

$$g(\beta) = \frac{1}{\beta_2 - \beta_1}, \quad \beta_1 \leq \beta \leq \beta_2. \tag{11.19}$$

By combining the likelihood function (11.16) with the product of (11.18) by (11.19), the joint posterior density for F_0 and β, given all data $\mathbf{ad}$ (i.e., the warranty data for the past product plus the early failure data for the new product), follows:

$$\pi(F_0, \beta | \mathbf{ad}) = \frac{1}{D} \prod_{k=1}^{l} [(1 - F_0)^{(x_{k-1}/\bar{x}_0)^\beta} - (1 - F_0)^{(x_k/\bar{x}_0)^\beta}]^{m_k} (1 - F_0)^{X(\beta)+b-1} F_0^{a-1},$$
$$\tag{11.20}$$

where the dependence on past data is implicitly conveyed by the beta parameters a and b (see relations (11.6)). The normalizing constant D is:

$$D = \int_0^1 \int_{\beta_1}^{\beta_2} \prod_{k=1}^{l} \left[(1 - F_0)^{(x_{k-1}/\bar{x}_0)^\beta} - (1 - F_0)^{(x_k/\bar{x}_0)^\beta} \right]^{m_k} (1 - F_0)^{X(\beta)+b-1} F_0^{a-1} \, d\beta \, dF_0. \tag{11.21}$$

Thus, the marginal posterior density $\pi(F_0 | \mathbf{ad})$ for the unreliability of the new product at the average mileage $\bar{x}_0$ is obtained by (numerically) integrating the joint density (11.20) over β:

$$\pi(F_0 | \mathbf{ad}) = \int_{\beta_1}^{\beta_2} \pi(F_0, \beta | \mathbf{ad}) \, d\beta. \tag{11.22}$$

By considering the approximation (11.14) between p_0 and F_0, the posterior density (11.22) can also be viewed as the posterior density $\pi(p_0|\mathbf{ad})$ for the failure probability p_0 of the new product during the warranty period t_0, given the warranty data for the past product, the judgements of the designers, and the early failure data for the new product.

A γ probability limit on p_0, say p_γ, can be obtained from the posterior density (11.22) by iteratively solving:

$$\int_0^{p_\gamma} \pi(p_0|\mathbf{ad})\, dp_0 = \gamma .\qquad(11.23)$$

11.4.2 Posterior Prediction of the Number of Failed Items

A posterior prediction of the number $m(t_0)$ of new-product items that will fail during the warranty period among the future population of N vehicles can also be made. By integrating the product of the predicted likelihood (11.9) and the posterior density (11.22), the posterior predictive distribution of $m(t_0)$, given all data $\mathbf{ad}$, results in:

$$P[m(t_0) = m|\mathbf{ad}] = \int_0^1 P[m(t_0) = m|p_0]\pi(p_0|\mathbf{ad})\, dp_0 .\qquad(11.24)$$

From (11.24), a conservative γ posterior prediction limit on $m(t_0)$, say $m_\gamma(t_0|\mathbf{ad})$, can also be obtained.

11.5 A Case Study

The case study described here involves predicting the reliability of a newly revised simple subsystem assembled in a car model that was already on the market. All data have been slightly modified for the sake of industrial confidentiality.

The automotive manufacturer intends to estimate the probability that the upgraded version of the outsourced component C will fail up to a warranty period of $t_0 = 3$ years. The upgraded component will be mounted on different car models that will presumably operate under different working conditions.

Estimates of the failure probability $\bar{p}_{0,i}$ of each part i ($i = 1,\dots,11$) of the original (past) version of the component C, defined as the fraction of failures that occurred during the warranty period over the population of the past component, are given in Table 11.3. The failure frequency of the past component as a whole, say $\bar{p}_0 = 0.01665$, is judged to be too high, and so design and process modifications are planned for some parts of the past component in an attempt to reduce the failure frequency of the upgraded version.

The designer indicates a lower limit $L_{0,i}$ for the failure probability of each modified part, taking into account the planned modifications; if a part is not modified, then $L_{0,i}$ is set equal to $\bar{p}_{0,i}$. These limits are predicted, for the moment, with the designer being unaware of any possible changes in the working conditions of the upgraded component and/or potential requests to reduce supplier costs. Then, for each modified part, the designer quantifies his uncertainty over the ability to actually attain the planned improvement by assigning a value to the "uncertainty factor" ρ_i. On the basis of this information, an upper limit $U_{0,i}$ for the failure probability of each part of the upgraded component is evaluated accordingly to (11.1). The predicted mean and standard deviation of the upgraded component, under unchanged conditions, are then:

$$E[p_0|\mathbf{pd}] = \sum_i (U_{0,i} + L_{0,i})/2 = 0.010474$$

$$\sigma[p_0|\mathbf{pd}] = \sqrt{\sum_i (U_{0,i} - L_{0,i})^2/12} = 0.001171 \qquad (11.25)$$

Table 11.3 provides all of the information formulated by the designer regarding the expected effectiveness of the planned modifications.

For a given car model M that will include the upgraded component C, the working conditions are known to be slightly more deleterious than those associated with the past component, so the "working factors" δ_W and η_W are set equal to $\delta_W = 1.2$ and $\eta_W = 2.0$ (see Table 11.1).

Moreover, the automotive manufacturer has requested a moderate cost reduction for the component C from the supplier, so the "cost factors" are $\delta_C = 1.5$ and $\eta_C = 1.5$ (as suggested in Table 11.2). Thus, from (11.4), the predicted mean and standard deviation of the failure probability p_0 of the new product included in car model M are, respectively:

$$E[p_0|\mathbf{pd}] = 0.01885 \quad \text{and} \quad \sigma[p_0|\mathbf{pd}] = 0.003513. \qquad (11.26)$$

Once the prior mean and standard deviation of p_0 have been predicted, the parameters of the beta prior density (11.5) on p_0 are found to be $a = 28.232$ and $b = 1469.2$. Then, from (11.9), the $\gamma = 0.90$ upper limit on p_0 is $p_{0.90} = 0.2347$.

On the basis of the informative prior density for p_0, the 0.05 and 0.95 conservative limits for the number $m(t_0)$ of items of the upgraded component that will fail during the warranty period in a future population of $N = 200,000$ vehicles are $m_{0.05}(t_0|\mathbf{pd}) = 2688$ and $m_{0.95}(t_0|\mathbf{pd}) = 5000$, respectively (see also the first line of Table 11.4).

Table 11.4 Prior and posterior inference on the failure probability and the number of future failures of the new product

	$E[p_0|\bullet]$	$\sigma[p_0|\bullet]$	$p_{0.90}$	$m_{0.05}(t_0|\bullet)$	$m_{0.95}(t_0|\bullet)$
Informative prior	0.01885	0.003513	0.02347	2688	5000
Informative posterior	0.01931	0.003245	0.02348	2839	4976
Vague posterior	0.02117	0.009047	0.03305	1935	7643

The predicted values of p_0 and $m(t_0)$ allow the management to decide that, considering both the more deleterious working conditions and the moderate cost reduction, the planned modifications are satisfactory and no further improvements are needed before starting mass production.

One year after mass production of the upgraded component has begun, the reliability team updates the reliability estimate on the basis of early field failure warranty data. These data refer to 18,000 vehicles with the upgraded version of the component C that were sold during the last 12 months.

A total of 18 failures occurred. On the basis of a log-normal distribution for the vehicle mileage (with a mean equal to 15,000 km per year and a standard deviation of 7000 km per year) and for the age of the vehicle, the number of vehicles for which the component has not failed is distributed in $l = 9$ mileage bands each with a width of 3000 km, as shown in Table 11.5. Likewise, the 18 failures are assigned to the mileage bands. Table 11.6 gives, for each mileage band, the number m_k of failures, the number c_k of suspended vehicles, and the number r_k of vehicles at risk.

By using the recursive formula (11.17), a nonparametric estimate for the reliability of the new component is obtained and plotted on Weibull paper. As depicted in Fig. 11.4, the two-parameter Weibull model seems to adequately fit the observed failure data, and the absence of infant mortality appears plausible.

The beta prior density for p_0 (with $a = 28.232$ and $b = 1469.2$) is then used in conjunction with the uniform prior density (11.19) for β over the interval $(1, 5)$. Note that this choice does not make use of the β estimate available from the Weibull plot. Thus, the "informative" marginal posterior density for p_0 is obtained from (11.20), and the mean and standard deviation for this, respectively, are:

$$E[p_0|\mathbf{ad}] = 0.01921 \quad \text{and} \quad \sigma[p_0|\mathbf{ad}] = 0.003245 . \tag{11.27}$$

From (11.23), the 0.90 posterior probability limit for p_0 results in $p_{0.90} = 0.02348$.

Table 11.5 Number of vehicles for which the component has not failed, distributed into mileage bands (in km)

Mileage band		Vehicle age [in months]											
x_{k-1}	x_k	1	2	3	4	5	6	7	8	9	10	11	12
0	3000	619	1175	796	510	205	59	21	7	3	1	1	0
3000	6000	9	379	1027	1604	1264	632	361	190	125	69	44	25
6000	9000	0	21	157	567	780	595	486	342	281	188	144	89
9000	12,000	0	1	24	136	283	300	320	282	284	235	216	165
12,000	15,000	0	0	4	31	94	123	162	168	196	186	192	164
15,000	18,000	0	0	1	10	32	52	74	86	113	119	133	124
18,000	21,000	0	0	0	2	9	17	34	44	63	71	85	86
21,000	24,000	0	0	0	1	3	7	13	20	32	42	53	56
24,000	27,000	0	0	0	0	1	2	6	9	16	22	30	35
27,000	∞	0	0	0	0	1	2	4	8	16	23	37	46
Total		628	1576	2009	2861	2672	1789	1481	1156	1129	956	935	790

Table 11.6 Field warranty data for the new component

Mileage band		No. of failures	No. of suspended vehicles	No. of vehicles at risk
x_{k-1}	x_k	m_k	c_k	r_k
0	3000	1	3397	18,000
3000	6000	5	5729	14,602
6000	9000	2	3650	8868
9000	12,000	3	2246	5216
12,000	15,000	1	1320	2967
15,000	18,000	3	744	1646
18,000	21,000	2	411	899
21,000	24,000	0	227	486
24,000	27,000	1	121	259
27,000	∞	0	137	137

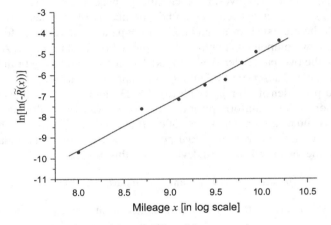

Fig. 11.4 Nonparametric estimate of the reliability of the new product

On the basis of the "informative" marginal posterior density for p_0, the 0.05 and 0.95 posterior limits for the number $m(t_0)$ of failed items among the future population of 200,000 vehicles are, respectively, $m_{0.05}(t_0|\mathbf{ad}) = 2839$ and $m_{0.95}(t_0|\mathbf{ad}) = 4976$ (see also the second line of Table 11.4).

By comparing prior and posterior results, we note that the use of early failure warranty data for the new product in conjunction with prior information on β yields:

1. A slight reduction in the uncertainty over the failure probability p_0
2. A confirmation of the value of the 0.90 upper limit on p_0
3. A reduction in the uncertainty over the number of failed items among the future population of upgraded components.

For comparative purposes, the reliability characteristics of the upgraded component are also estimated under the assumption that prior information on the failure prob-

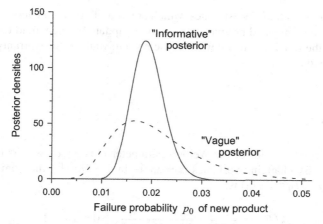

Fig. 11.5 "Informative" (——) and "vague" (– – –) posterior densities of the failure probability of the new product

ability of the new product cannot be elicited (i.e., no past data are available), and only the uniform prior density (11.19) of β over the interval $(1, 5)$ is used.

The last line of Table 11.4 provides the inferential results obtained by using the resulting vague prior $g_V(F_0, \beta)$ (see the Appendix). Upon comparing the "informative" and "vague" estimation results, it appears that the use of the informative prior density of p_0 sensibly reduces the uncertainty over the reliability of the new product, both in terms of the failure probability (the standard deviation of p_0 is reduced by a factor of three), and in terms of the number of failed items among the future population of vehicles (the width of the 0.90 predictive interval on $m(t_0)$ is halved).

The gain provided by the technological information on p_0 is clearly shown by Fig. 11.5, where the "informative" and "vague" posterior densities of p_0 are compared.

11.6 Conclusions

The proposed procedure has been found to be suitable for making inferences on the reliability of innovated automotive components when technical knowledge about the design modifications is actually available or the designers and technologists can be asked about qualitative factors.

This methodology is very useful for exploiting reliability predictions in all the main phases of the product development process: from the very beginning, when the product and process design choices must be validated, to much later on, when commercialization has begun and the actual reliability as perceived by the customers must be certified.

Moreover, the reliability estimates, which are initially based on very limited failure data for the innovated product, can easily be updated as new field data become available to the reliability team, without the need to wait for the warranty period of each vehicle to elapse.

Appendix

The noninformative prior of the Weibull parameter α is $g_N(\alpha) \propto 1/\alpha$ (Evans and Nigm 1980). By changing variables $\alpha = x_0/[-\ln(1-F_0)^{1/\beta}]$, the noninformative prior density of $F_0 \equiv F(\bar{x}_0)$ is:

$$g_N(F_0) \propto -\frac{1}{\beta(1-F_0)\ln(1-F_0)}, \quad 0 \le F_0 \le 1. \tag{11.28}$$

By using density (11.28) as the prior of the failure probability p_0 together with the uniform prior (11.19) of β, the vague joint prior density follows (assuming that p_0 and β are prior independent random variables):

$$g_V(F_0, \beta) = g_N(F_0) \cdot g(\beta) \propto \frac{1}{\beta(1-F_0)\ln(1-F_0)}, \quad 0 \le F_0 \le 1, \ \beta_1 \le \beta \le \beta_2. \tag{11.29}$$

The marginal posterior density of p_0, given the new data **nd**, results in:

$$\pi_V(F_0|\mathbf{ad}) = \frac{1}{D_V} \int_{\beta_1}^{\beta_2} \frac{\prod_{k=1}^{l}[(1-F_0)^{(x_{k-1}/\bar{x}_0)^\beta} - (1-F_0)^{(x_k/\bar{x}_0)^\beta}]^{m_k}}{\beta(1-F_0)^{1-X(\beta)}\ln(1-F_0)} \, d\beta, \tag{11.30}$$

where the normalizing constant D_V is:

$$D_V = \int_0^1 \int_{\beta_1}^{\beta_2} \frac{\prod_{k=1}^{l}[(1-F_0)^{(x_{k-1}/\bar{x}_0)^\beta} - (1-F_0)^{(x_k/\bar{x}_0)^\beta}]^{m_k}}{\beta(1-F_0)^{1-X(\beta)}\ln(1-F_0)} \, d\beta \, dF_0. \tag{11.31}$$

Acknowledgements The authors would like to acknowledge Dr. Mario Vianello, formerly with Fiat Auto, for providing fundamental suggestions regarding the expert-opinion elicitation process.

References

Attardi, L., Guida, M., Pulcini, G.: A mixed-Weibull regression model for the analysis of automotive warranty data. Reliab. Eng. Syst. Safe. **87**, 265–273 (2005)

Campean, I.F., Kuhn, F.P., Khan, M.K.: Reliability analysis of automotive field failure warranty data. In: Zio, E., Demichela, M., Piccinini, N. (eds.): Safety and Reliability: Towards a Safer World. Politecnico di Torino, Torino, pp. 1337–1344 (2001)

Evans, I.G., Nigm, A.H.M.: Bayesian 1-sample prediction for the 2-parameter Weibull distribution. IEEE Trans. Reliab. **R-29**, 410–413 (1980)

Grohowski, G., Hausman, W.C., Lamberson, L.R.: A Bayesian statistical inference approach to automotive reliability estimation. J. Qual. Technol. **8**, 197–208 (1976)

Guida, M., Pulcini, G.: Automotive reliability inference based on past data and technical knowledge. Reliab. Eng. Syst. Safe. **76**, 129–137 (2002)

Guida, M., Pulcini, G.: Early reliability testing of automotive components by using past data and technical information. In: Proc. Int. Conf. on Degradation, Damage, Fatigue and Accelerated Life Models in Reliability Testing, Angers, France, pp. 54–61, 22–24 May (2006)

Guida, M., Pulcini, G., Vianello, M.: Early inference on reliability of upgraded automotive components by using past data and technical information. J. Stat. Plan. Infer. (in press) (2008), (doi: 10.106/j.jspi.2007.08.008)

Lu, M.-W.: Automotive reliability prediction based on early field failure warranty data. Qual. Reliab. Eng. Int. **14**, 103–108 (1998)

Lu, M.-W., Rudy, R.J.: Reliability test target development. In: Proc. Annu. Reliability and Maintainability Symp., Los Angeles, USA, pp. 77–81, 24–27 Jan. (2000)

Nelson, W.: Applied Life Data Analysis. Wiley, New York (1982)

Rai, B., Singh, N.: A modeling framework for assessing the impact of new time/mileage warranty limits on the number and cost of automotive warranty claims. Reliab. Eng. Syst. Safe. **88**, 157–169 (2005)

Vianello, M.: A Bayesian reliability estimate in the automotive context (in Italian). Associazione Tecnica dell'Automobile **34**, 618–624 (1981)

Chapter 12
Stochastic Processes for Modeling the Wear of Marine Engine Cylinder Liners

Massimiliano Giorgio, Maurizio Guida, and Gianpaolo Pulcini

Abstract In this chapter, a stochastic process-based approach is adopted to formulate the reliability function for cylinder liners of diesel engines used for marine propulsion, which fail when their wear exceeds a specified limit. In order to describe the wear process, three different stochastic models are proposed. The first two are based on age-dependent processes, namely a shock model with independent nonstationary increments and a gamma process. The third model is based on a state-dependent homogeneous Markov chain. All of these models have been applied to the analysis of a real case study relating to the cylinder liners of some diesel engines used in ships of the Grimaldi Group.

12.1 Introduction

The standard (statistical) approach to reliability modeling consists of selecting a model from a limited list of candidate parametric models and determining its unknown parameters. The candidates are usually prepacked black-box models. The model is selected via goodness of fit tests and/or based on considerations concerning the ability of the model to capture the characteristics of the aging of the system under

Massimiliano Giorgio
Department of Aerospatial and Mechanical Engineering, Second University of Naples
81031 Aversa (NA), Italy
e-mail: massimiliano.giorgio@unina2.it

Maurizio Guida
Department of Information Engineering and Electrical Engineering, University of Salerno
84084 Fisciano (SA), Italy
e-mail: mguida@unisa.it

Gianpaolo Pulcini
Istituto Motori, National Research Council, 80125 Napoli, Italy
e-mail: g.pulcini@im.cnr.it

Pasquale Erto, *Statistics for Innovation*
ISBN 978-88-470-0814-4, © Springer 2009

study. The values of the unknown parameters are determined using well-established estimation procedures. Both of these tasks are usually performed on the basis of failure data only.

This approach is not always fruitful or, at least, it is not always the most effective one. The most frequent reasons for this inadequacy are a scarcity of failure data, and the fact that the standard approach does not allow the direct use of other information and experimental data that are often available in practice. In addition, depending on the context, the reliability model often needs to have specific features that standard black-box models do not possess.

In these cases it may be convenient to adopt alternative solutions which can capitalize on all other available pieces of information to allow the formulation of ad hoc (customized) models. Over the last few decades, increasing effort has been devoted to the formulation of reliability models obtained by stochastic modeling of the failure-causing mechanism and/or the dependence of the system lifetime on endogenous and/or exogenous factors (see Singpurwalla 1995 for an excellent overview of this kind of model).

In this chapter, a stochastic process-based approach is adopted in order to formulate the reliability function of cylinder liners of diesel engines used for marine propulsion, which are wearing components that fail when the wear level reaches a specified limit.

This reliability function is formulated through the following two-step approach:

- Describe the wear process via a monotonically increasing stochastic process
- Obtain the reliability function from the distribution of the first (and sole) passage time to the wear limit

Three models are discussed in what follows. These models differ in the stochastic process adopted in order to describe the wear process. The first two are based on age-dependent processes, namely a shock model with independent nonstationary increments and a gamma process, respectively. The third model is based on a state-dependent homogeneous Markov chain.

None of these models represent a fully precise description of the failure-causing mechanism, however. Indeed, this mechanism is not completely known even to experts of diesel engines. In addition, the available experimental data do not enable one to perform the empirical calibration and validation of models based on highly complex hypotheses, since only measures of the wear level at the moments at which periodic inspections are performed are available. Thus, any model of practical utility must possess a simple mathematical structure, and must be based on a few, clear hypotheses, with the aim of capturing the leading features of the underlying wear mechanism. In accordance with this principle, the first model uses a nonhomogeneous Poisson shock arrival process and treats the effects of traumatic events as being unknown but deterministic. Thus, this model accounts for the stochastic nature of shock arrivals but does not account for the randomness of shock effects. The second model describes the wear process through a gamma process. The gamma process can be considered a shock model where the shocks occur randomly in a Poisson process; this model also has gamma-distributed shock effects, where the Poisson rate tends to infinity while the sizes of the effects tend to zero in proportion (Lawless et al. 2004). Thus, this model puts the focus directly on the distribution of cumulative wear increments, but does not explicitly account for the rate of occurrence of shocks. The third model describes the wear process through a state-dependent homogeneous Markov chain. Thus, it assumes that the wear growth in a future time interval depends on the present state of the liner and that, given the state, this growth is independent of the age of the liner.

All of the abovementioned models are simple and present a small number of parameters. Moreover, all of them treat the dependence of the residual life of a liner on its wear status in an explicit way. These features produce the following advantages:

- They allow one to simply implement all of the necessary inferential procedures on the basis of data that are actually available in practice, i.e., wear data. In order to avoid costly engine failures, the wear status of a liner is periodically observed during specific inspections, and a liner is replaced when the measured wear is close to a limit set beforehand. As a collateral effect, this effective condition-based maintenance strategy provides a large amount of wear data but strongly limits the size of lifetime samples.
- They enable one to easily perform wear-based predictions of the residual life of a liner. This feature becomes very useful when condition-based maintenance is of concern. In fact, the crucial issue when developing this maintenance strategy is to identify the replacement time that enables one to avoid both operational failures and unnecessary or premature preventive actions.

All of the proposed models have been applied to the analysis of a real wear data set relating to cylinder liners of diesel engines used by ships of the Grimaldi Group. Model parameters were estimated by ad hoc estimation procedures. Moreover, the accumulation of wear after pre-fixed time intervals was predicted given the present age or status of the liner. Finally, the residual life distribution function and the mean residual life were estimated. A comparison of the results provided by the three proposed models closes the chapter.

12.2 The Case Study: System Description, Technological Information and Experimental Data

One of the main factors in the failure of heavy-duty diesel engines used for marine propulsion is the cylinder liner wear (see, e.g., Kodali et al. 2000; Li et al. 2002; Gadow et al. 2003). In such engines, wear usually occurs in the top region of the liner (the top death center, TDC), where the maximum mechanical and thermal load occurs (Li et al. 2002). Many studies agree that in this region:

- Wear is mainly due to the presence on the piston surface of abrasive particles produced by the combustion of heavy fuels and oil degradation (soot)

Table 12.1 Wear data for 30 cylinder liners (t [hours]; w [mm])

i	n_i	$t_{i,1}$	$w_{i,1}$	$t_{i,2}$	$w_{i,2}$	$t_{i,3}$	$w_{i,3}$	$t_{i,4}$	$w_{i,4}$
1	3	11,300	0.90	14,680	1.30	31,270	2.85		
2	2	11,360	0.80	17,200	1.35				
3	2	11,300	1.50	21,970	2.00				
4	2	12,300	1.00	16,300	1.35				
5	3	14,810	1.90	18,700	2.25	28,000	2.75		
6	3	9700	1.10	19,710	2.60	30,450	3.00		
7	3	10,000	1.20	30,450	2.75	37,310	3.05		
8	3	6860	0.50	17,200	1.45	24,710	2.15		
9	3	2040	0.40	12,580	2.00	16,620	2.35		
10	3	7540	0.50	8840	1.10	9770	1.15		
11	3	8510	0.80	14,930	1.45	21,560	1.90		
12	4	18,320	2.20	25,310	3.00	37,310	3.70	45,000	3.95
13	3	10,000	2.10	16,620	2.75	30,000	3.60		
14	2	9350	0.85						
15	1	13,200	2.00						
16	1	7700	1.05						
17	1	7700	1.60						
18	1	8250	0.90						
19	1	3900	1.15						
20	1	7700	1.20						
21	1	9096	0.50						
22	1	19,800	1.60						
23	1	10,450	0.40						
24	1	12,100	1.00						
25	3	12,000	1.95	27,300	2.70	49,500	3.15		
26	1	8800	1.40						
27	1	2200	0.40						
28	2	33,000	2.90	38,500	3.25				
29	2	8800	0.50	27,500	2.15				
30	1	8250	0.70						

- Wear is also due to the action of corrosive agents, identified as sulfuric acid, nitrous/nitric acids and water

As a consequence, according to the experts, the wear at a given age t can be viewed as a nondecreasing cumulative damage process.

Wear data for this case study (see Table 12.1) refer to 30 cylinder liners of diesel engines used in ships of the Grimaldi Group.

The wear was measured during periodic inspections by an ad hoc measurement instrument with a sensitivity of 0.05 mm. Thus, the data consist of a number n_i ($i = 1, \ldots, 30$) of inspections performed on each liner, the age $t_{i,j}$ ($j = 1, \ldots, n_i$) [in operating hours] of the liner at the moment of inspection, and the accumulated wear $W(t_{i,j})$ [in mm] observed during each inspection.

Figure 12.1 shows the wear paths of the 30 liners under study (the measured points of each liner have been linearly interpolated in this graphical presentation). The behavior of the wear paths is in accordance with results reported in the lit-

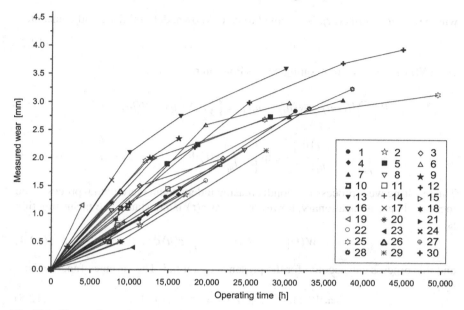

Fig. 12.1 Observed wear processes

erature. In fact, the observed (interpolated) processes are closely interwoven, thus indicating the nondeterministic nature of the wear process. Moreover, as is usually the case for these kinds of paths (see, for example, Gertsbakh et al. 1969), it is possible to discern an early phase (accommodation period) where the rate of wear decreases noticeably and a subsequent zone with a quasi-constant rate.

We note that, in the present application, the accommodation phase occupies almost the entire observed portion of each liner's lifetime, since warranty clauses specify that the liner should be substituted before it accumulates a rather "low" wear level (4 mm), in order to avoid catastrophic and very expensive failures.

12.3 Formulating the Models

12.3.1 Model 1: Shock Model with a Deterministic and Constant Effect

The wear process is described by the following shock model (Esary and Marshall 1973):

$$W(t) = \sum_{i=1}^{N(t)} Y_i \,, \tag{12.1}$$

where the shock effect, Y_i, is assumed to be unknown, deterministic and constant:

$$P[Y_i = c] = 1, \quad \forall i = 1, 2, \ldots, N(t), \tag{12.2}$$

and $\{N(t); t \geq 0\}$ is a nonhomogeneous Poisson process:

$$P\left[\bigcap_{i=1}^{k} [N(u_i) - N(h_i) = n_i]\right] = \prod_{i=1}^{k} P[N(u_i) - N(h_i) = n_i]$$

$$= \prod_{i=1}^{k} \frac{\left[\int_{h_i}^{u_i} z(t)\,dt\right]^{n_i}}{n_i!} \exp\left[-\int_{h_i}^{u_i} z(t)\,dt\right], \quad h_i < u_i \leq h_{i+1} \tag{12.3}$$

The resulting wear process is a nondecreasing independent incremental process. The increments are nonstationary, however, unless $z(t)$ is constant. The mean function is:

$$E[W(t)] \equiv E[cN(t)] = c \int_0^t z(u)\,dz, \tag{12.4}$$

and this process can account for the presence of a trend. The variance function is:

$$\text{Var}[W(t)] \equiv \text{Var}[cN(t)] = c^2 \int_0^t z(u)\,dz, \tag{12.5}$$

which is a nondecreasing function. Since:

$$P[W(t) = ck] = P[N(t) = k] = \frac{\left[\int_0^t z(u)\,du\right]^k}{k!} \exp\left[-\int_0^t z(u)\,du\right], \tag{12.6}$$

the reliability function can be expressed as:

$$R(t) \equiv P[T > t] = P[W(t) < w_{max}] = P[N(t) < k_{max}]$$

$$= \sum_{k=0}^{k_{max}-1} \frac{\left[\int_0^t z(u)\,du\right]^k}{k!} \exp\left[-\int_0^t z(u)\,du\right] \tag{12.7}$$

where T is the liner lifetime, w_{max} is the pre-fixed wear limit, and $k_{max} = \text{int}[w_{max}/c]$ is the smallest integer greater than or equal to w_{max}/c.

The residual life distribution, given the liner's status at time t, results in:

$$P[T - t \leq \tau | W(t) = w_t] = P[W(t + \tau) \geq w_{max} | W(t) = w_t]$$

$$= \sum_{k=\Delta k_{max}}^{\infty} \frac{\left[\int_t^{t+\tau} z(u)\,du\right]^k}{k!} \cdot \exp\left[-\int_t^{t+\tau} z(u)\,du\right]; \quad \tau > 0, \ w_t < w_{max} \tag{12.8}$$

where $\Delta k_{max} = \text{int}[(w_{max} - w_t)/c]$.

This model was used to describe the wear process of the abovementioned cylinder liners on the basis of a reduced data set in Bocchetti et al. (2006) and Giorgio et al. (2007) by assuming both a log-linear mean function:

$$M(t) \equiv E[N(t)] = (a/b)[\exp(bt) - 1], \tag{12.9}$$

and a power-law (PL) mean function:

$$M(t) \equiv E[N(t)] = \left(\frac{t}{a}\right)^b .$$
(12.10)

Note that the mean function (12.9) is bounded. In contrast, function (12.10) is unbounded. For many degradation processes the use of a bounded mean function may be appropriate. Nevertheless, the use of a bounded mean function can produce estimates for the residual life that are too optimistic, especially when very old components that are hardly degraded are of interest.

This issue is not discussed any further in this chapter. Thus, in the following, only the more conservative model (12.10) is considered, termed the "PL shock model" here.

The mean and variance functions for the PL shock model are:

$$E[W(t)] = c\left(\frac{t}{a}\right)^b \quad \text{and} \quad \text{Var}[W(t)] = c^2\left(\frac{t}{a}\right)^b .$$
(12.11)

12.3.2 Model 2: Gamma Wear Process

The gamma wear process (Cinlar 1980; Dufresne et al. 1991; Van Noortwijk et al. 2004) is formulated as follows:

$$P\left[\bigcap_{i=1}^{k}[W(u_i) - W(h_i) > w_i]\right] = \prod_{i=1}^{k}P[W(u_i) - W(h_i) > w_i]$$

$$= \prod_{i=1}^{k}\int_{w_i}^{\infty}\frac{w^{v(u_i)-v(h_i)-1}}{\Gamma[v(u_i) - v(h_i)]\eta^{v(u_i)-v(h_i)}}\exp(-w/\eta)\,dw, \quad h_i < u_i \le h_{i+1}, \quad (12.12)$$

where $v(t)$ is a nondecreasing function defined for $t \ge 0$ with $v(0) = 0$ and $\eta > 0$.

Thus, the gamma process has gamma-distributed independent increments. These increments are nonstationary, unless $v(t)$ is a linear function of t. The mean function of the gamma wear process is:

$$E[W(t)] = \eta v(t) ,$$
(12.13)

and it can be nonlinear; thus, this model can account for the presence of a trend.

The variance:

$$\text{Var}[W(t)] = \eta^2 v(t) ,$$
(12.14)

is a nondecreasing function of t. The probability density function of $W(t)$ is:

$$f_{W(t)}(w) = \frac{w^{v(t)-1}}{\Gamma[v(t)]\eta^{v(t)}}\exp(-w/\eta) ,$$
(12.15)

and the reliability function can be expressed as:

$$R(t) \equiv P[T > t] = P[W(t) < w_{max}] = \int_0^{w_{max}} \frac{w^{v(t)-1}}{\Gamma[v(t)]\eta^{v(t)}} \exp(-w/\eta)\,dw. \quad (12.16)$$

The residual life distribution, given the status at time t, results in:

$$P[T - t \leq \tau | W(t) = w_t] = P[W(t + \tau) \geq w_{max} | W(t) = w_t]$$

$$= \int_{w_{max}-w_t}^{\infty} \frac{w^{v(t+\tau)-v(t)-1}}{\Gamma[v(t+\tau) - v(t)]\eta^{v(t+\tau)-v(t)}} \exp(-w/\eta)\,dw, \quad \tau > 0;\ w_t < w_{max}.$$

$$(12.17)$$

A typical choice for $v(t)$ is the power law function:

$$v(t) = \left(\frac{t}{\alpha}\right)^{\beta}, \quad (12.18)$$

and the resulting mean and variance functions are, respectively:

$$E[W(t)] = \eta \left(\frac{t}{\alpha}\right)^{\beta} \quad \text{and} \quad \text{Var}[W(t)] = \eta^2 \left(\frac{t}{\alpha}\right)^{\beta}. \quad (12.19)$$

This specific gamma process has been widely applied. As an example, in Elling-wood et al. (1993) the values $\beta = 1$, $\beta = 2$ and $\beta = 0.5$ were used to describe the degradation of concrete due to the corrosion of reinforcement, sulfate attack and diffusion-controlled aging, respectively. In Cinlar et al. (1977), $\beta = 1/8$ was adopted to model creep propagation. Finally, in Hoffmans et al. (1995) and Van Noortwijk et al. (1999), $\beta = 0.4$ was used to model degradation due to erosion (expected scour-hole depth).

12.3.3 Model 3: State-Dependent Markov Wear Process

This model assumes that $W(t)$ is a homogeneous, monotonically increasing Markov process. Thus, the wear increment $dW(t) = W(t + dt) - W(t)$ in the interval $(t, t + dt)$ depends on the wear level at time t, $W(t)$, while, due to homogeneity, and given $W(t)$, it does not depend on t. In addition, the expected value of $dW(t)$ is assumed to decrease exponentially with $W(t)$:

$$E[dW(t)] \propto \exp[-BW(t)]. \quad (12.20)$$

Then a discrete-time approximation is made by dividing the time axis into contiguous and equally spaced intervals of length h. Under such an approximation, $W(t_m)$ denotes the wear level at time $t_m = m \cdot h$, with $W(0)$ being the initial state at $t_0 = 0$.

Further, a discrete-state approximation is made by considering a nondecreasing Markov chain with countable state space $(0, \Delta w, 2\Delta w, \ldots)$, where Δw is a constant that represents the sensitivity of the measuring instrument (namely, $\Delta w = 0.05$ mm).

Using such approximations, the probability that the wear level at the time $t_m = m \cdot h$ is equal to $W(t_m) = r_m \cdot \Delta w$, given that the wear level $W(t_{m-1}) = r_{m-1} \cdot \Delta w$ at the time $t_{m-1} = (m-1) \cdot h$ is assumed to be a Poisson probability:

$$
\begin{aligned}
P_{r_{m-1}, r_m} &\equiv P[W(t_m) = r_m \cdot \Delta w | W(t_{m-1}) = r_{m-1} \cdot \Delta w] \\
&= \begin{cases} \dfrac{[A \exp(-Br_{m-1}\Delta w)]^{r_m - r_{m-1}}}{(r_m - r_{m-1})!} \exp[-A \exp(-Br_{m-1}\Delta w)], & r_m \geq r_{m-1} \\ 0, & r_m < r_{m-1} \end{cases}
\end{aligned}
$$
$$(12.21)$$

where $A \cdot \exp(-B \cdot r_{m-1} \cdot \Delta w) = A \cdot \exp[-B \cdot W(t_{m-1})]$ is the expected wear increment in the elementary interval $[(m-1) \cdot h, m \cdot h]$, given the state $W(t_{m-1})$ at the time t_{m-1}. This Markov process will henceforth be called the Poisson Markov process.

Thus, the probability that the wear level at the time $t_m = m \cdot h$ is equal to $W(t_m) = r_m \cdot \Delta w$, given the initial state $W(0) = r_0 \cdot \Delta w$ at the time $t_0 = 0$, can be calculated as:

$$
\mathbf{P}[W(t_m) = r_m \cdot \Delta w | W(0) = r_0 \cdot \Delta w] = [\mathbf{P}^m]_{r_0, r_m}, \tag{12.22}
$$

where $\mathbf{P}$ is the one-step transition matrix of the Markov chain, whose elements are given by Eq. 12.21.

The reliability function can be expressed as:

$$
\mathbf{P}[W(t_m) < r_{max} \cdot \Delta w | W(0) = r_0 \cdot \Delta w] = \sum_{r=0}^{r_{max}-1} [\mathbf{P}^m]_{r_0, r}, \tag{12.23}
$$

where $r_{max} = w_{max}/\Delta w$ (note that, since Δw is the measuring instrument sensitivity, both r_{max} and any $r_i = W(t_i)/\Delta w$ are integers).

Finally, the residual life distribution, given the wear status $W(t_i) = r_i \cdot \Delta w$, at time $t_i = i \cdot h$, results in:

$$
\begin{aligned}
&P[T - t_i \leq \tau | W(t_i) = r_i \cdot w_i] \\
&= P[W(t_i + \tau) \geq w_{max} | W(t_i) = r_i \cdot w_i] = \sum_{r=\Delta r_{max}}^{\infty} [\mathbf{P}^m]_{r_0, r}, \quad \tau > 0, \; w_i < w_{max},
\end{aligned}
$$
$$(12.24)$$

where $\Delta r_{max} = r_{max} - r_i$.

Given the initial state $W(0) = r_0 \cdot \Delta w$, the mean and variance of $W(t_i)$ are, respectively:

$$
E[W(t_i) | W(0) = r_0 \cdot \Delta w] = \sum_{r=r_0}^{\infty} r\Delta w [\mathbf{P}^i]_{r_0, r}, \tag{12.25}
$$

and:

$$\text{Var}[W(t_i)|W(t_0) = r_0 \cdot \Delta w]$$

$$= \sum_{r=r_0}^{\infty} \{r \cdot \Delta w - E[W(t_i)|W(t_0) = r_0 \cdot \Delta w]\}^2 [\mathbf{P}^i]_{r_0,r}. \qquad (12.26)$$

This model has already been used to describe the degradation process of catalytic converters in Barone et al. (2008).

12.4 Numerical Application and Results

All of the proposed models have been applied to the wear data of Table 12.1. Their unknown parameters were estimated by the procedures described in the Appendix, and the following numerical values were obtained:

PL shock model: $\hat{a} = 660.0\,\text{h}, \hat{b} = 0.7244, \hat{c} = 0.1661\,\text{mm},$
Gamma process: $\hat{\alpha} = 611.6\,\text{h}, \hat{\beta} = 0.7351, \hat{\eta} = 0.1513\,\text{mm},$
Poisson Markov process: $\hat{A} = 1.334 \cdot 10^{-4}\,\text{mm}, \hat{B} = 0.2696\,\text{mm}^{-1}.$

Based on these estimates, the mean wear functions for the three models were computed. These curves are plotted in Fig. 12.2 along with the "observed" average wear at selected times, obtained by averaging over the lines which interpolate the measured wear levels of each liner. It appears that all of the proposed models adequately fit the observed data, since the mean wear curves follow the "observed" trend.

In particular, note that the curves of the age-dependent models (namely, the PL shock model and the gamma process) are very close to each other.

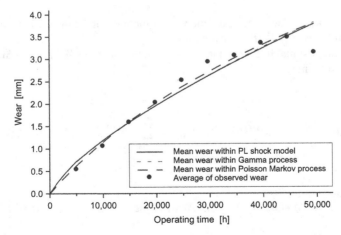

Fig. 12.2 Average observed wear (•) and estimates of the expected values of $W(t)$

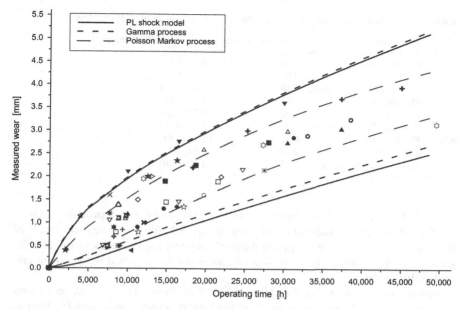

Fig. 12.3 Estimated 90% probability intervals for $W(t)$ according to the proposed wear models

Figure 12.3 shows the approximate 90% probability intervals for $W(t)$ according to the proposed models. The band for each age-dependent model includes a fraction of the observed wear points which is very close to the exact probability that a point will occur inside the band: 5 points out of a total of 58 fall outside the 90% probability band. On the other hand, the 90% probability band for the state-dependent model appears to be too narrow, since it includes only 73% of the observed wear points.

This can be explained by the fact that the Poisson Markov process is indexed by only two parameters (instead of three parameters, as in the other models), which means that this model is not able to accurately describe both the behavior of the mean curve and the behavior of the process variance. Thus, further studies are now exploring more flexible forms of transition probabilities by relaxing the Poisson distribution (12.21) and adopting models indexed by a larger number of parameters.

For a number of liners, namely the liners labeled 8, 13, and 23, we estimated the wear growth after a time interval $\Delta t = 10,000$ hours, given the present age t_{i,n_i} or the present wear $W(t_{i,n_i})$ of the selected liner. These liners were chosen because:

- Liner 8 accumulated a wear of 2.15 mm up to $t_{8,3} = 24,710$ hours, which is close to the estimated expected wear $E[W(24,710)] \cong 2.30$ mm,
- Liner 13 accumulated a wear of 3.60 mm up to $t_{13,3} = 30,000$ hours, which is much larger than the estimated expected wear $E[W(30,000)] \cong 2.70$ mm,
- Liner 23 accumulated a wear of 0.40 mm up to $t_{23,1} = 10,450$ hours, which is much smaller than the estimated expected wear $E[W(10,450)] \cong 1.20$ mm.

Table 12.2 Predicted wear growth (in mm) after a future time interval of 10,000 hours

Liner	Wear model	Expected growth	90% Probability interval
#8	PL shock model	0.64	$(0.17, 1.16)$
	Gamma process	0.65	$(0.23, 1.25)$
	Poisson Markov process	0.68	$(0.40, 0.95)$
#13	PL shock model	0.61	$(0.17, 1.16)$
	Gamma process	0.62	$(0.22, 1.20)$
	Poisson Markov process	0.48	$(0.25, 0.70)$
#23	PL shock model	0.77	$(0.17, 1.33)$
	Gamma process	0.78	$(0.31, 1.41)$
	Poisson Markov process	1.05	$(0.70, 1.40)$

Table 12.2 gives both the expected growth $E[\Delta W]$ and the 90% probability interval for ΔW for each combination of liner and model. We first note that, for all of the liners analyzed, the age-dependent models, namely the PL shock model and the gamma process, provide fairly similar estimates for the expected wear growth and probability intervals.

In addition, the estimates of $E[\Delta W]$ for liner 8 are close to each other, because its present level of wear is close to the expected wear of a liner that is as old as liner 8. For liner 13, the age-dependent models overestimate the wear growth compared to the state-dependent model, because this liner has much more wear than a liner with the same age, and so its wear growth in the future is expected to be smaller in a state-dependent model than in an age-dependent model.

On the other hand, the age-dependent models underestimate the wear growth of liner 23 with respect to the state-dependent model, because this liner has much less wear than would be expected of a liner of its age. Hence, its future wear growth is expected to be larger in a state-dependent model than in an age-dependent model.

Finally, the Poisson Markov process provides probability intervals that are much narrower than those provided by the other models. This can be explained by the fact that this process is not able to describe the whole process variance.

Figures 12.4–12.6 depict the estimates for the residual reliability of the selected liners according to the proposed models. Table 12.3 shows the estimates for the mean residual lifetimes (corresponding to the areas under each residual reliability curve) of the selected liners. These results are in full agreement with those of Table 12.2.

Table 12.3 Estimated residual lifetimes (in hours) of the liners labeled 8, 13, and 23

Wear model	Liner 8	Liner 13	Liner 23
PL shock model	34,192	8179	60,232
Gamma process	32,100	7614	59,163
Poisson Markov process	32,850	9371	51,270

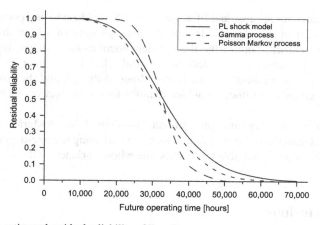

Fig. 12.4 The estimated residual reliability of liner 8

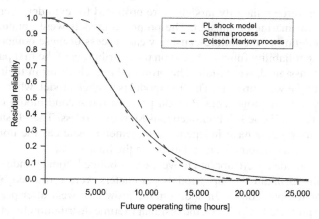

Fig. 12.5 The estimated residual reliability of liner 13

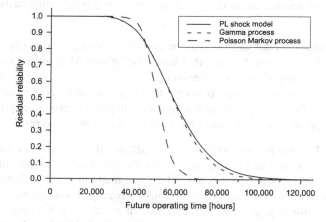

Fig. 12.6 The estimated residual reliability of liner 23

The estimates for the residual lifetime of liner 8 are close to each other, because the present wear of this liner is close to the average wear of a liner that is as old as liner 8. In contrast, both of the the age-dependent models underestimate (compared to the state-dependent model) the residual reliability of liner 13, because it has accumulated more wear than an average liner of its age, and they overestimate the residual reliability of liner 23, which exhibits less wear than would be expected at its age.

The residual reliability functions estimated using the Poisson Markov process are less dispersed than the reliability functions estimated using the age-dependent models, since this model is not able to describe the whole variance of the wear process.

12.5 Conclusions

In this chapter, three reliability models were proposed for cylinder liners of diesel engines used for marine propulsion. The proposed models were obtained through stochastic modeling of the failure-causing wear process of these components, and by deriving the reliability function based on the distribution of the first passage time to a fixed, pre-assigned, wear limit. The proposed models differ in their stochastic descriptions of the wear process. The first model is a shock model, where the shocks are driven by a nonhomogeneous Poisson process; the second model is a gamma process, and the third one is a homogeneous Markov process. The PL shock model and the gamma process have independent increments, whereas the homogeneous Markov process has increments that depend on the liner's state.

All of these models were applied to wear data obtained from cylinder liners used in ships of the Grimaldi Group. Models parameters were estimated by adopting ad hoc estimation procedures. In addition, the growth in wear after pre-fixed time intervals was predicted. Moreover, the residual lifetime distribution function and the mean residual lifetime were estimated.

Based on the results obtained, it is possible to draw the following two main conclusions:

1. The PL shock model and the gamma process (which provide very similar results) fit the data well. On the other hand, the Poisson Markov process does not adequately model the process variance.
2. When the wear of the liner is very different from the level of wear expected for a liner of its age, age-dependent and state-dependent processes provide very different predictions, in terms of both future wear growth and residual life.

The drawback of the Poisson Markov model alluded to in (1) is not a serious one, since it can be overcome by adopting a more flexible Markov model, indexed for example by three or more parameters. The point made in (2) is a slightly trickier issue. Indeed, it shows that the predictions made strongly depend on the model used, so model selection appears to be a crucial task. In general, it is not easy to select an adequate model. The selection is often based on goodness-of-fit tests and on ex post analyses. In many cases, however, this statistical approach gives poor results due to the inadequacy or scarcity of experimental data. Hence, when possible, model selection should be preferably guided by physical or technical knowledge rather than statistical procedures.

For the application presented in the chapter, technical information suggests that a state-dependent model is more appropriate than an independent increment model. Thus, further studies are now in progress to formulate a more flexible state-dependent Markov model and to assess its performance in terms of goodness-of-fit and prediction ability.

A Appendix: Estimation Procedures

A.1 PL Shock Model

The PL shock model is indexed by three parameters, namely the PL parameters a and b and the elementary wear c. The likelihood function based on the observed wear data $W(t_{i,j})$ $(i = 1, \ldots, m; \ j = 1, \ldots, n_i)$ is:

$$L(a,b,c) = \left\{ \prod_{i=1}^{m} \prod_{j=1}^{n_i} \frac{[(t_{i,j}/a)^b - (t_{i,j-1}/a)^b]^{W_{i,j}/c}}{(W_{i,j}/c)!} \right\} \exp\left[-\sum_{i=1}^{m} (t_{i,n_i}/a)^b \right] , \quad (12.27)$$

where $W_{i,j} = W(t_{i,j}) - W(t_{i,j-1})$ is the wear accumulated by liner i during the inspection interval $(t_{i,j-1}, t_{i,j})$ (with $W(t_{i,0}) \equiv 0$ and $t_{i,0} \equiv 0$), and t_{i,n_i} is the age of liner i at the time of the last inspection. The log-likelihood function can then be written as:

$$\ell(a,b,c) = \sum_{i=1}^{m} \sum_{j=1}^{n_i} \frac{W_{i,j}}{c} \left[\ln\left(t_{i,j}^b - t_{i,j-1}^b \right) - b\ln(a) \right] - \sum_{i=1}^{m} \sum_{j=1}^{n_i} [\ln(W_{i,j}/c)!]$$
$$- \sum_{i=1}^{m} (t_{i,j}/a)^b . \quad (12.28)$$

Unfortunately, both a^b and c act as multiplicative factors for the cumulated wear, and consequently the maximum likelihood estimation (MLE) procedure is not able to estimate a^b and c separately. Thus, to overcome this estimation problem, the following five-step estimation procedure, which combines the maximum likelihood method with the first moment estimation method, is proposed (see Giorgio et al. 2007 for a detailed description):

1. Obtain the MLE of b by numerically solving the equation:

$$\sum_{i=1}^{m} \sum_{j=1}^{n_i} W_{i,j} \frac{t_{i,j}^b \ln(t_{i,j}) - t_{i,j-1}^b \ln(t_{i,j-1})}{t_{i,j}^b - t_{i,j-1}^b} - \sum_{i=1}^{m} W(t_{i,n_i}) \frac{\sum_{i=1}^{m} t_{i,n_i}^b \ln(t_{i,n_i})}{\sum_{i=1}^{m} t_{i,n_i}^b} = 0 , \quad (12.29)$$

where $W_{i,j} = W(t_{i,j}) - W(t_{i,j-1})$, and $W(t_{i,n_i})$ is the wear accumulated by the liner i up to the last inspection.

2. Obtain the MLE of $\theta = c/a^b$:

$$\hat{\theta} = \frac{\sum_{i=1}^{m} W(t_{i,n_i})}{\sum_{i=1}^{m} t_{i,n_i}^{\hat{b}}} . \tag{12.30}$$

3. Estimate the expected wear accumulated by the liner i during $(t_{i,j-1}, t_{i,j})$:

$$\hat{E}[W_{i,j}] = \hat{\theta}\left(t_{i,j}^{\hat{b}} - t_{i,j-1}^{\hat{b}}\right), \quad i = 1,\ldots,m, \ j = 1,\ldots,n_i , \tag{12.31}$$

and compute:

$$\hat{y}_k = \frac{W_{i,j} - \hat{E}[W_{i,j}]}{\sqrt{\hat{E}[W_{i,j}]}}, \quad k = 1,\ldots,N = \sum_{i=1}^{m} n_i , \tag{12.32}$$

where $k = j + \sum_{l=0}^{i-1} n_l$.

4. Estimate the elementary wear c:

$$\hat{c} = \text{Var}[\hat{y}_k] = \frac{\sum_{k=1}^{N} (\hat{y}_k - \text{Ave}[\hat{y}_k])^2}{N-1} , \tag{12.33}$$

where $\text{Ave}[\hat{y}_k] = \sum_{k=1}^{N} \hat{y}_k/N$ is the sample mean of $\hat{y}_k$.

5. Estimate the PL parameter a:

$$\hat{a} = (\hat{c}/\hat{\theta})^{1/\hat{b}} . \tag{12.34}$$

Note that the accuracy of the estimate of c, and hence of a, depends on the accuracy of $\hat{E}[W_{i,j}]$ and on the size N of the sample of $\hat{y}_k$. Clearly, a larger N leads to a more accurate estimate for c. If the sample mean of $\hat{y}_k$ is close to 0, then we can be confident that the estimate of c is accurate. In particular, in the present case study, the sample mean of $\hat{y}_k$ is equal to 0.0243. This value is much less than the sample standard deviation $\sigma(\hat{y}_k) = 0.1661^{0.5} = 0.408$, making us confident that the estimate of c is sufficiently accurate.

A.2 Gamma Process

Under the assumption of a gamma process, the likelihood function of the observed data is:

$$L(\alpha, \beta, \eta) = \left(\prod_{i=1}^{m} \prod_{j=1}^{n_i} \frac{W_{i,j}^{(t_{i,j}/\alpha)^\beta - (t_{i,j-1}/\alpha)^\beta - 1}}{\eta^{(t_{i,j}/\alpha)^\beta - (t_{i,j-1}/\alpha)^\beta} \Gamma[(t_{i,j}/\alpha)^\beta - (t_{i,j-1}/\alpha)^\beta]} \right)$$

$$\times \exp\left[-\sum_{i=1}^{m} \sum_{j=1}^{n_i} W_{i,j}/\eta \right] \tag{12.35}$$

and the ML estimates for α, β, and η can be obtained by maximizing the logarithm of the likelihood $L(\alpha,\beta,\eta)$ through a numerical optimization procedure.

A.3 Poisson Markov Process

Assuming a Poisson Markov process, the likelihood function based on the observed data $r_{i,j} = W_{i,j}/\Delta$ $(i = 1,\dots,m;\ j = 1,\dots,n_i)$ is given by:

$$L(A,B) = \prod_{i=1}^{m}\prod_{j=1}^{n_i} [\mathbf{P}^{m_{i,j}-m_{i,j-1}}]_{r_{i,j-1},r_{i,j}} \, , \qquad (12.36)$$

where $m_{i,j} = \text{int}[t_{i,j}/h + 0.5]$. The model parameters A and B are then estimated by maximizing the logarithm of the likelihood $L(A,B)$ through a numerical optimization procedure.

References

Barone, S., Giorgio, M., Guida, M., Pulcini, G.: Stochastic modeling and prediction of catalytic converters degradation. In: Martorell, S., Guedes Soares, C., Barnett, J. (eds.): Safety, Reliability and Risk Analysis. CRC Press, London, pp. 2251–2258 (2009)

Bocchetti, D., Giorgio, M., Guida, M., Pulcini, G.: A competing risk model for the reliability of cylinder liners in ship diesel engines. In: Guedes Soares, C., Zio, E. (eds.): Safety and Reliability for Managing Risk. Taylor & Francis, London, pp. 2751–2759 (2006)

Cinlar, E., Bazant, Z.P., Osman, E.: Stochastic process for extrapolating concrete creep. J. Eng. Mech. Div. **103**, 1069–1088 (1977)

Cinlar, E.: On a generalization of Gamma processes. J. Appl. Prob. **17**(2), 467–480 (1980)

Dufresne, F., Gerber, H.U., Shiu, E.S.W.: Risk theory with the gamma process. ASTIN Bull. **21**, 177–192 (1991)

Ellingwood, B.R, Mori, Y.: Probabilistic methods for condition assessment and life prediction of concrete structures in nuclear power plants. Nucl. Eng. Des. **142**, 155–66 (1993)

Esary, J.D., Marshall, A.W.: Shock models and wear processes. Ann. Probab. **1**, 627–649 (1973)

Gadow, R., Lopez, D., Perez, J.: Carbon deposition reducing coatings for highly loaded large diesel engines. In: Proc. 2003 SAE World Congr., Detroit, MI, USA, SAE Paper 2003-01-1100, 3–6 March (2003)

Gertsbakh, I.B., Kordonskiy, K.B.: Models of Failures. Springer-Verlag, Berlin (1969)

Giorgio, M., Guida, M., Pulcini, G.: A wear model for reliability assessment of cylinder liners in marine diesel engines. IEEE Trans. Reliab. **56**, 158–166 (2007)

Hoffmans, G.J.C.M., Pilarczyk, K.W.: Local scour downstream of hydraulic structures. J. Hydraul. Eng. **121**, 326–340 (1995)

Kodali, P., How, P., McNulty, W.D.: Methods of improving cylinder liner wear. In: Proc. SAE 2000 World Congr., Detroit, MI, USA, SAE Paper 2000-01-0926, 6–9 March (2000)

Lawless, J.F., Crowder, M.J.: Covariates and random effects in a gamma process model with application to degradation and failure. Lifetime Data Anal. **10**, 213–227 (2004)

Li, S., Csontos, A.A., Gable, B.M., Passut, C.A., Jao, T.-C.: Wear in Cummins M-11/EGR test engine. Proc. SAE Int. Fuels & Lubricants Meeting & Exhibition, Reno, USA, SAE Paper 2002-01-1672, 6–9 May (2002)

Singpurwalla, N.D.: Survival in dynamic environments. Stat. Sci. **10**, 86–103 (1995)

Van Noortwijk, J.M., Klatter, H.E.: Optimal inspection decisions for the block mats of the Eastern-Scheldt barrier. Reliab. Eng. Syst. Safe. **65**, 203–211 (1999)

Van Noortwijk, J.M., Pandey, M.D.: A stochastic deterioration process for time-dependent reliability analysis. In: Maes, M.A., Huyse, L. (eds.): Reliability and Optimization of Structural Systems. Taylor & Francis, London, pp. 259–265 (2004)

Part IV
Research and Innovation Management

Chapter 13
A New Control Chart Achieved via Innovation Process Approach

Pasquale Erto and Giuliana Pallotta

Abstract An alternative title for this chapter could be "Innovation for Statistics", which would be equally appropriate to illustrate its contents. In fact, the chapter shows how to perform the innovation process approach in order to fulfill a scientific research need in a specific application area. More specifically, this chapter describes the incremental development of a new control chart of the Weibull percentile (i.e. the reliable life) as a practical example of a product attained following an innovation process approach. More specifically, the design of the chart required three "short" operative steps of innovation process, in order to provide the original chart with new peculiar features incrementally.

13.1 Introduction

The proposed control chart is a case study of a new product attained following an incremental innovation process approach.

It is well known that the Statistical Process Control (SPC) methods are extensively used in several contexts in order to monitor and improve the quality of production processes and service operations. Within this framework, the relevance of common control charts is recognized but, obviously, do not cover all the potential needs and/or application areas.

Pasquale Erto
Department of Aerospace Engineering, University of Naples Federico II
P. le Tecchio 80, 80125, Naples, Italy
e-mail: pasquale.erto@unina.it

Giuliana Pallotta
Department of Aerospace Engineering, University of Naples Federico II
P. le Tecchio 80, 80125, Naples, Italy
e-mail: g.pallotta@unina.it

Pasquale Erto, *Statistics for Innovation*
ISBN 978-88-470-0814-4, © Springer 2009

In particular, dealing with manufacturing industry, any technological product must ensure a specified reliability level, otherwise the company will incur high costs due to defective items, that include costs of reworking and/or scraping, costs of returns, costs of failures in-service, warranty replacements and the loss of good-will, hardly restorable. Once the reliability target levels have been achieved by means of a good design and tests, an outstanding need is monitoring the reliability performance of the item, launched on the market. This needed monitoring is generally performed through a control chart in order to detect unusual reliability variation due to "assignable causes" and to adopt the necessary corrective actions as soon as possible.

Nearly always, the distributions of the reliability parameters are skewed and the classical Shewhart control charts are not effective, due to the frequent small size of the available samples that prevent one to use the so called "normalizing effect". Unfortunately, very few papers dealing with non-normal populations and reliability control can be found in literature such as (Padgett and Spurrier 1990), (Kanji and Arif 2001), (Xie et al. 2002), (Shore 2004), (Zhang and Chen 2004), (Nichols and Padgett 2005).

In this context, many new needs can be identified. Among them, we found the following one: the transformation of also non-statistical data (i.e. data not in a quantitative statistical form) into useful information for reliability control. In fact, the possibility of exploiting also non-statistical data provides the analyst with a competitive advantage, allowing him to face situations where the data are scarce and/or the responsiveness of the chart is crucial. Consequently, the analyst is primarily concerned with the identification of non-statistical data sources and then with the generation of useful information from them.

In the following sections we will list out the "short" operative steps of the innovation process which led to the development of the new chart.

13.2 Identifying the New Needed Estimation Features

The reliability control based on pure classical estimation procedures is carried out ignoring a great amount of non-statistical data, usually available as prior technological knowledge in engineering field. On the other hand the frequent scarcity of statistical reliability data (caused by the low number of the items in operation, the very high level of their reliability, etc.) makes the use of non "classical" reliability estimators competitive.

Facing these new needs, in order to innovate the estimation process we can use more effective (non "classical") reliability estimators like those based on the application of the Bayes theorem, which allows one to merge statistical and technological data. In this way, the estimators could provide the estimation process with more efficiency not wasting any kind of data achieved by usual engineering design and process control practice.

We found adequate using specific Bayesian reliability estimators known as "Practical Bayes Estimators" (PBE) since they were developed from an Engineers' point of view and applied in several technological contexts during past years (Erto 1982); (Erto 2005); (Erto and Lanzotti 1991); (Erto and Giorgio 1996); (Erto and Giorgio 2000). They reinforce the statistical experimental information by easily incorporating the available Engineers' prior technological knowledge into the estimation procedure.

Differently from other Bayesian reliability estimators, the PBE require only the specification of those parameters, of the "prior distributions", into which the Engineers' prior technological knowledge can really be converted, leaving the remaining parameters unspecified. Moreover, in (Erto 2005) we can find two practical examples proving the effectiveness of PBE compared to the Maximum Likelihood ones, when few statistical data are available but a significant amount of technological data is available.

Acronyms and Notation

Sf$\{\cdot\}$	survival function

pdf$\{\cdot\}$	probability density function
R	reliability level
K	constant equal to $\ln(1/R)$
x_R	equivalent to the $1-R$ percentile of the Weibull distribution such as that Sf$\{x_R\} = R$
x	data set
n	sample size
k	order number of the current sample (initial value $k = 1$)
α	tail area of a pdf
δ, β	scale and shape parameters of the Weibull distribution
β_1, β_2	prior numerical interval for β
$\wedge$	implies an estimate
PBE	Practical Bayes Estimators (or Estimates)
L	Likelihood

Among all the distributions used in technology, the Weibull distribution is the most widely used as a reliability model. The Weibull survival function is:

$$\text{Sf}\{x; \delta, \beta\} = \exp\left[-(x/\delta)^\beta\right]; \quad x \geq 0; \ \delta, \beta > 0 \qquad (13.1)$$

being δ and β the scale and shape parameters of the Weibull distribution respectively.

Since the available (non-statistical) Engineers' information can be more easily converted in terms of reliable life x_R and shape parameter β, around these two parameters the whole estimation process has to be centred. Consequently, we must reparameterize the (13.1) in terms of these parameters:

$$\text{Sf}\{x; x_R, \beta\} = \exp\left[-K\left(x/x_R\right)^{\beta}\right] \; ; \quad x \geq 0; \; x_R, \beta > 0$$

$$K = \ln\left(1/R\right) \tag{13.2}$$

x_R and n both being unknown.

As a matter of fact, in reliability problems, the control of the Weibull percentile x_R is strategic. The choice of x_R is justified by the wide use of this parameter in:

- defining warranty conditions;
- developing contractual specifications;
- expressing key-indicators in engineering catalogues.

In practical situations the control of a specific Weibull percentile becomes crucial when the quality characteristic of interest is the breaking strength of brittle materials (such as carbon and boron) or the tensile adhesive strength (Park 1996). In these contexts, monitoring process mean and variance (by means of classical control charts) could be seriously less effective than monitoring a specific percentile since a small variation in mean and/or variance can produce a significant shift in the percentile of interest (Padgett and Spurrier 1990).

Moreover, in life tests, the Weibull percentile measures the reliable life x_R useful to evaluate the performance of mechanical devices. In this applicative field, we must outline that, if the reliability level R of the tested items is very high, we are able to collect very few data which prevent us to use classical control charts. In these cases, the proposed Bayesian approach may result an appropriate approach.

In engineering, very often some knowledge exists about the mechanism of failure under consideration, which can be converted into quantitative statistical form about β. Moreover, an engineer presumably knows more than the simple order of magnitude of the life which the designed item has, e.g. he has a quite precise knowledge about x_R. Then, with both these pieces of information, he can formulate a numerical interval (β_1, β_2) for β and an anticipated value for x_R. The PBE adopt some specific priors which are suitable to express this kind of prior technological knowledge about x_R and β and, moreover, are tractable too. For a comprehensive and detailed discussion see (Erto 1982).

Combining the two priors for x_R and β by multiplying the conditional prior for x_R and the prior for β, we can obtain a joint prior probability density function for x_R and β:

$$\text{pdf}\{x_R, \beta\} = \text{pdf}\{x_R|\beta\} \times \text{pdf}\{\beta\} \; . \tag{13.3}$$

Then the joint prior (13.3) can be integrated with the experimental data, usually available as a sample array of n data, resulting from reliability tests.

So, we can use the Bayes theorem that substantially says:

$$\begin{pmatrix} \text{joint posterior} \\ \text{of unknown } x_R \text{ and } \beta \end{pmatrix} \propto \begin{pmatrix} \text{joint prior} \\ \text{of } x_R \text{ and } \beta \end{pmatrix} \times \begin{pmatrix} \text{likelihood function} \\ \text{of the sample} \end{pmatrix}$$

to obtain a joint posterior probability density function for x_R and β. "Prior" and "posterior" mean before and after obtaining experimental data respectively. So, in this way, the theorem fuses the technological prior knowledge, summarized into the joint prior, with all the information (data and shape of the reliability model) included into the likelihood function.

We can say that the joint posterior density function describes the residual uncertainty which exists about the two parameters. So we could estimate the parameters x_R and β adopting their modal or median or mean values. The PBE choose the last ones, that is, the expectations of x_R and β.

It is interesting to note that using a random variable transformation, the joint posterior probability density function for x_R and β can be transformed into a standard Gamma. This property can be used to calculate the percentiles of the marginal posterior probability density function for x_R as those needed to draw an x_R control chart, the control limits being simple transformations of the percentiles of the standard Gamma. For technical details see Appendix B in (Erto and Pallotta 2006).

13.3 Developing the Innovative Estimation Procedure

In the following subsections we will present the innovation steps which led to the development of the new chart. We report an incremental analysis of its innovative features, gradually examined.

13.3.1 First Innovation Step: A Shewhart-Type Control Chart of the Weibull Percentile

First we developed a Shewhart-type control chart of the Weibull percentile, which can be easily drawn thanks to its similarity to the common Shewhart charts.

Following a classical approach, the chart works as a "single sample" scheme since it only uses the information about the process contained in the last plotted point. As shown in Fig. 13.1, the estimation procedure processes samples one by one, being the single plotted point representative of *the current sample information combined with the initial prior information.*

Moreover this "first step" chart is *static* since its design parameters (sample size n, sampling interval, control limits) are left unchanged.

Thanks to the Bayesian nature of our estimation process, we can use all the available data to gain useful information about future values of x_R. Thus, on the basis of the overall posterior distribution of x_R, derived applying the PBE to the whole collected dataset, we are able to anticipate a credibility interval for the parameter x_R. In this way, we can predict that our x_R estimates will likely and fairly vary within the anticipated range if no significant change in reliability happens. Therefore, our control chart plays a predictive role very similar to that of all the Shewhart-type

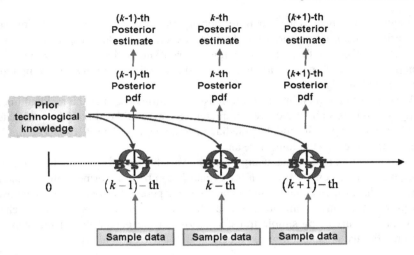

Fig. 13.1 The working engine of a "single sample" estimation procedure

control charts. For a comprehensive discussion see (Erto and Pallotta 2007). For a description of the chart as an example of a Data Technology product see (Erto et al. 2008).

The following real applicative example, using in control and out-of-control datasets proposed in (Padgett and Spurrier 1990) respectively reported in Table 13.1 and Table 13.2, is presented to show the operative use of the chart.

The control of the first percentile (equivalent to the reliable life x_R) of the break-ing strength distribution of a certain type of carbon fibers is considered. This kind of carbon fiber is produced to manufacture composite materials which need fibers with a breaking strength greater than 1.22 GPa (Giga-Pascal) with 99% probability.

With the objective to control that this specification ($x_R = 1.22$, $R = 0.99$) is met, a sample of $n = 5$ fibers, each 50 mm long, is selected from the manufacturing pro-cess periodically, and the breaking stress of each fiber is measured.

When the process was certainly in-control, ten samples of size $n = 5$ of breaking stress of these carbon fibers were sampled (see Table 13.1). On the basis of some past knowledge and experience, the Weibull distribution is considered to closely fit such breaking stress and is assumed to be the underlying distribution of the control chart (being the assumption also supported by some goodness-of-fit tests for the Weibull distribution, not reported here).

Figure 13.2 shows the control chart of the first percentile estimated by means of the PBE, depicting the sequence of x_R estimates in time domain. The chart is drawn using the data of Table 13.1. On the basis of the information provided in (Padgett and Spurrier 1990), the prior interval $(2.8, 6.8)$ for β and the anticipated (mean) value 1.22 for x_R are adopted. Applying the PBE estimators to the whole data set (50 data), the posterior estimate for β is 4.69 and the posterior estimate for x_R is 1.20. So, the central line (CL) is 1.20; given $\alpha = 0.27\%$, the lower control limit (LCL) and the upper control limit (UCL), are respectively 0.97 and 1.67.

Table 13.1 Breaking stresses (GPa) of carbon fibers
(in control state, $x_R = 1.22$, $R = 0.99$)

k-th sample	Stress				
1	3.70	2.74	2.73	2.50	3.60
2	3.11	3.27	2.87	1.47	3.11
3	4.42	2.41	3.19	3.22	1.69
4	3.28	3.09	1.87	3.15	4.90
5	3.75	2.43	2.95	2.97	3.39
6	2.96	2.53	2.67	2.93	3.22
7	3.39	2.81	4.20	3.33	2.55
8	3.31	3.31	2.85	2.56	3.56
9	3.15	2.35	2.55	2.59	2.38
10	2.81	2.77	2.17	2.83	1.92

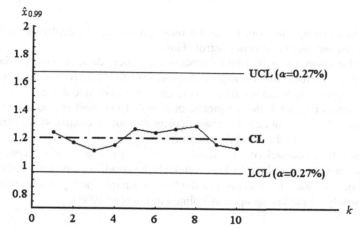

Fig. 13.2 Shewhart-type control chart of the Weibull percentile (x_R, $R = 0.99$) using the in-control process data (Table 13.1) with $n = 5$

Using always the same prior interval $(2.8, 6.8)$ for β and the anticipated (mean) value 1.22 for x_R, the ordinates of the ten points (i.e. the estimates of the breaking strength) corresponding to the ten samples of size $n = 5$ are calculated.

As we can see, the iterative application of the estimation process results to be practicable and viable, since we can combine the same prior technological information (achieved from non-statistical data) together with the statistical data available at each sampling step.

In order to show the responsiveness of the chart, we can consider the ten samples in Table 13.2. Differently from previous samples, they are representatives of an out-of-control state, simulated by shifting the first percentile to 0.26 GPa from the original value of 1.22 GPa. It is interesting to note that this shift could really be originated by a decrease in the mean and an increase in the variance of the breaking stress and expresses a likely process deterioration. As anticipated in Sect. 13.2, this

Table 13.2 Breaking stresses (GPa) of carbon fibers
(out-of-control state, $x_R = 0.26$, $R = 0.99$)

k-th sample	Stress				
1	1.41	3.68	2.97	1.36	0.98
2	2.76	4.91	3.68	1.84	1.59
3	3.19	1.57	0.81	5.56	1.73
4	1.59	2.00	1.22	1.12	1.71
5	2.17	1.17	5.08	2.48	1.18
6	3.51	2.17	1.69	1.25	4.38
7	1.84	0.39	3.68	2.48	0.85
8	1.61	2.79	4.70	2.03	1.80
9	1.57	1.08	2.03	1.61	2.12
10	1.89	2.88	2.82	2.05	3.65

simultaneous change turn out to be the most critical out-of-control state, hardly
detectable by means of classical control charts.

Then, the above chart is used to process these new data. As before, the prior
interval $(2.8, 6.8)$ for β and the anticipated (mean) value 1.22 for x_R are adopted, in
order to calculate the estimates from all the ten samples of size $n = 5$.

As shown in Fig. 13.3, the diagnostic property of the chart is good. The graphical position of the points on the chart depicts the out-of-control state, caused by
a significant change in the reliability parameters.

However, the drawback of this Shewhart-type control chart is not keeping memory of the past sampled data. In fact, although the results of previous samples are
recorded on the chart, none is used in the "single sample" scheme whose decision
depends solely on the last sample, as pointed out in (Lai 1995).

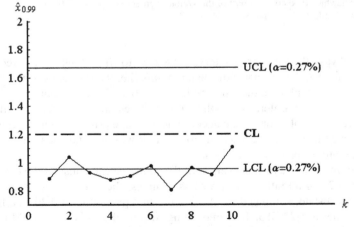

Fig. 13.3 Shewhart-type control chart of the first Weibull percentile (x_R, $R = 0.99$) using the out-
of-control process data (Table 13.1) with $n = 5$

This feature causes a relative insensitiveness of Shewhart-type control charts to moderate shifts in the process, which results to be particularly critical for reliability control. So, a further innovation step is motivated.

13.3.2 Second Innovation Step: A Bayesian Cumulative Control Chart of the Weibull Percentile

In order to improve the responsiveness of control chart, another feature has been added to the proposed chart, starting from the opportunity provided by the Bayes Theorem to continually update the knowledge about the parameters of interest as soon as a new sample becomes available. Thus, we used the PBE to perform a recursive estimation procedure which is the core of the proposed Bayesian cumulative control chart.

As a matter of fact, this "second step" chart shows a great similarity to the CUSUM charts since it incorporates past data into the estimation procedure. Page in (Page 1954) first proposed the Cumulative Sum charts (commonly known as CUSUM charts) where current and past data are accumulated to effectively detect small shifts in the mean. In this way, he first fulfilled a need for new, more effective, control schemes, after the launch of the Shewhart control charts.

Even if it is similar to CUSUM charts, the proposed chart differs from them due to a new feature: it is able to combine both current and past data together with the technological knowledge about the process by applying recursively the Bayes Theorem sample by sample. Thus, exploiting the PBE under recursive use, each point, plotted on the chart, is representative of *the whole information accumulated until that moment* (as shown in Fig. 13.4).

In order to show its effectiveness, the cumulative estimation procedure is compared to a "single sample" procedure. In Fig. 13.5 the same data (reported in Table 13.1 and Table 13.2) are plotted collecting them in samples of size $n = 5$. The Bayesian cumulative sequence of points (formed of the ten in control points followed by the ten out-of-control ones) is graphically connected by a solid line. The corresponding "single sample" sequence is graphically connected by a dashed line. A vertical dashed line is traced in correspondence to the occurrence of the out-of-control state. The control limits are calculated pooling all the available in control data, leaving them unchanged. For a detailed description see (Erto and Pallotta 2006).

As we can see, the *cumulative* procedure results more effective than the "single sample" one and the resulting chart can successfully compete with the few known alternative ones (Padgett and Spurrier 1990), (Kanji and Arif 2001), (Xie et al. 2002), (Zhang and Chen 2004), (Nichols and Padgett 2005).

However this "second step" chart lacks the following feature, very appreciated in recent research in SPC, which could differentiate the chart from the *static* Shewhart-type control charts: the opportunity to update at least one of its design parameters (sample size n, sampling interval or control limits) in response to the current sample

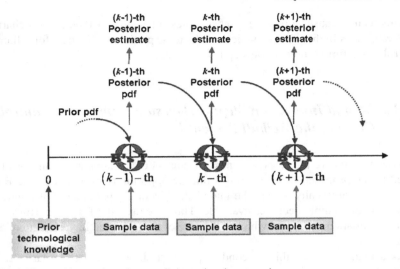

Fig. 13.4 The working engine of a *cumulative* estimation procedure

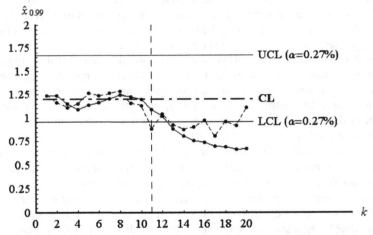

Fig. 13.5 Bayesian *cumulative* (solid line) and "single sample" (dashed line) sequence of points (x_R, $R = 0.99$) representatives of the 10 out-of-control samples (Table 13.1) which follow the 10 in control ones (Table 13.1) with $n = 5$

information. We found adequate to apply the cumulative estimation procedure, used to obtain the points to be plotted, to compute the control limits too. So, almost naturally, the subsequent research need turns out to be the design of adaptive control limits.

13.3.3 Third Innovation Step: A Bayesian Cumulative and Adaptive Control Chart of the Weibull Percentile

Thus, in order to fulfill the new identified need and to provide a more reliable picture of the process to be monitored, we adopted a control policy based on adaptive control limits.

For a conservative behavior of the chart, we set the following decision rule about the possibility to update the control limits:

$$\text{If } \hat{x}_{R,k} \in [\text{LCL}_{k-1}, \text{UCL}_{k-1}] \Rightarrow \text{ update: } \begin{cases} \text{LCL} = \text{LCL}_k \\ \text{UCL} = \text{UCL}_k \end{cases}$$

$$\text{If } \hat{x}_{R,k} \notin [\text{LCL}_{k-1}, \text{UCL}_{k-1}] \Rightarrow \text{ do not update: } \begin{cases} \text{LCL} = \text{LCL}_{k-1} \\ \text{UCL} = \text{UCL}_{k-1} \end{cases} \quad (13.4)$$

In this way, the rule aims at pointing out a critical situation, leaving the control limits unchanged if an out-of-control state is suspected. This empirical rule conforms to the human physiological behavior, which consists in reacting by freezing whenever a peril is felt.

The benefit of the procedure is clear: the adaptive estimation of the control limits allows the chart to choose the control policy which better fits the current situation. Thus the policy exploits the whole current state of knowledge, attaining the time-varying control limits as suggested in (Steiner 1999) and (Mastrangelo and Brown 2000).

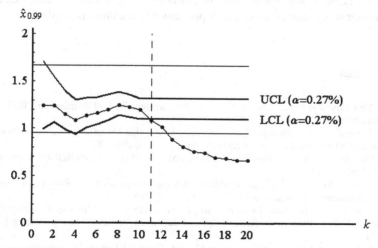

Fig. 13.6 Bayesian sequence of points $(x_R, R = 0.99)$ representatives of the 10 out-of-control samples (Table 13.2) which follow the 10 in control ones (Table 13.1) plotted against the adaptive (*boldface line*) control limits and unchanged ones with $n = 5$

In order to show the advantage deriving from adapting control limits over time, the same data (reported in Table 13.1 and Table 13.2) are plotted collecting them in samples of size $n = 5$ (Fig. 13.6) and updating the limits according to the proposed "physiological-type" control rule. Again, a vertical dashed line is traced in correspondence to the occurrence of the out-of-control state. As we can see the diagnostic properties of the chart improve and, thanks to their time-varying nature, more likely control limits, narrowed over time, result.

13.4 Conclusions

The scientific research which led to the new chart has been inspired to a continuous innovation process. At each operative step:

- firstly a new need in control charting is identified, a choice to fulfill the identified need is taken and a new feature is added to the chart;
- secondly, the deriving advantages are evaluated and, starting from the analysis of the drawbacks, a further need is identified and a new innovation step is stimulated.

Following this approach, the responsiveness of the chart has gradually but continuously improved. The chart can support the analyst in facing some critical scenarios where small samples are available, the values of the process parameters have to be estimated, being unknown, and very prompt decisions are needed.

Obviously, the attained result must be considered as a research process step itself, since further study about the statistical properties of the chart is needed.

References

Erto, P.: New Practical Bayes estimators for the 2-parameters Weibull distributions. IEEE Trans. on Reliability **31**(2), 194–197 (1982)

Erto, P.: Integration of technological and statistical knowledge for reliability control. Proc. 55th Session of International Statistical Institute, Sydney, Australia (2005)

Erto, P., Giorgio, M.: Modified Practical Bayes Estimators. IEEE Trans. on Reliability **45**(1), 132–137 (1996)

Erto, P., Giorgio, M.: Assessing high reliability via Bayesian approach and accelerated tests. Reliability Engineering & System Safety **76**(3), 303–312 (2002)

Erto, P., Lanzotti, A.: Bayesian Estimators for the parameters of the Inverse Weibull reliability model (in Italian), Proc. of the XXXVI Conf. of the Italian Statistical Society, Pescara **2**, 59–66 (1992)

Erto, P., Pallotta, G.: A Bayesian Cumulative Memory Control Chart for Weibull Processes. Proc. 9th Int. QMOD, Liverpool, UK, 300–310 (2006)

Erto, P., Pallotta, G.: A New Control Chart for Weibull Technological Processes. Quality Technology and Quantitative Management **4**(4), 553–567 (2007)

Erto, P., Pallotta, G., Park, S.H.: An Example of Data Technology Product: A Control Chart for Weibull Processes. International Statistical Review **76**(2), 157–166 (2008)

Johnson, L., Kotz, S., Balakrishnan, N.: Continuous univariate distributions, 1, Wiley, New York (1994)

Kanji, G.K., Arif, O.H.: Median rankit control chart for Weibull distribution. Total Quality Management 12(5), 629–642 (2001)

Lai, T.L.: Sequential changepoint detection in quality control and dynamical systems. J. of the Royal Statistical Society. Series B 57(4), 613–658 (1995)

Mastrangelo, C.M., Brown, E.C.: Shift Detection Properties of Monitoring Centerline Control Charts Schemes. J. of Quality Technology 32(1), 67–74 (2000)

Nichols, M.D., Padgett, W.J.: A Bootstrap Control Chart for Weibull Percentiles. Quality and Reliability Engineering Int. 22(2), 141–151 (2005)

Padgett, W.J., Spurrier, J.D.: Shewhart-type charts for percentiles of strength distributions. J. of Quality Technology 22(4), 283–288 (1990)

Page, E.S.: Continuous Inspection Schemes. Biometrika 41(1/2), 100–115 (1954)

Park, S.H.: Robust design and analysis for quality engineering. Chapman & Hall, London (1996)

Shore, H.: Non-normal populations in quality applications: a revisited perspective. Quality and Reliability Engineering Int. 20, 375–382 (2004)

Steiner, S.H.: EWMA control charts with time-varying control and fast Initial Response. J. of Quality Technology 31(1), 75–86 (1999)

Xie, M., Goh, T. N., Ranjan P.: Some effective control chart procedures for reliability monitoring. Reliability Engineering and System Safety 77(2), 143–150 (2002)

Zhang, L., Chen, G.: EWMA charts for monitoring the mean of censored Weibull lifetimes. J. of Quality Technology 36(3), 321–328 (2004)

Chapter 14
A Critical Review and Further Advances in Innovation Growth Models

Pasquale Erto and Amalia Vanacore

Abstract In recent decades, the literature on technology management has proposed *S*-curves as promising tools for analyzing the life-cycle of technological innovation in order to support company strategies and policies.

Nevertheless, the scant attention devoted to the analytical foundations of the *S*-curve model has limited its capacity to actually model the performance of technological innovation.

In this work, the main mathematical and statistical characteristics of the most popular *S*-curve models are analyzed in order to verify their suitability for modeling technological innovation processes. In particular, the properties of the linearity of each model have been studied, since they are needed in order to make some useful inferences.

Critical comparative analysis has been carried out using some real data sets for three different technologies: piston aero-engines, jet aero-engines, and digital signal processors.

14.1 Introduction

Innovation is any technological change that enables the ratio between the expected/obtained benefits and the investment required to be improved. In other words, a company innovates when it uses a technology—not necessary the most advanced one—to gain a competitive edge.

Pasquale Erto
Department of Aerospace Engineering, University of Naples Federico II
P. le Tecchio 80, 80125, Naples, Italy
e-mail: `pasquale.erto@unina.it`

Amalia Vanacore
Department of Ae rospace Engineering, University of Naples Federico II
P. le Tecchio 80, 80125, Naples, Italy
e-mail: `amalia.vanacore@unina.it`

Pasquale Erto, *Statistics for Innovation*
ISBN 978-88-470-0814-4, © Springer 2009

In order to completely manage technological processes, the company should rely on a model that allows the evolution of the performance of the technology to be analyzed. In the literature on technology management, the S-curve is a very popular (although often only qualitative) model used to analyze the innovation life-cycle. Proposed by Richard Foster (1986), it is based on a fairly straightforward concept: the curve tracks the growth in performance as a function of the effort devoted to developing the technology, or as an appropriate function of time.

The name of the model derives from the observation that the plot of performance growth is usually approximately S-shaped. In fact, at the beginning, the performance growth of any new technology is slow. Then, as a critical mass of technology expertise builds up, this growth becomes rapid. After a while, as the technology matures, the growth slows again until it reaches a saturation limit. Thus, starting from the curve's inflection point, ploughing further work into an old technology will result in diminishing returns (Asthana 1995). In the literature on technology management, the S curve model has been mainly used a posteriori to analyze the growth in the performance of a technology. Nevertheless, it is undoubtedly true that the S-curve model is best applied to a priori analysis. For example, it can be employed to optimize the productivity of a technology, forecast the evolution of the technology, and monitor the effects of management initiatives during the innovation process. However, it is also true that the proper use of the S-curve model as a useful strategic tool requires a deep knowledge of its genesis as well as of its analytical aspects.

The contribution presented in this chapter deals with the main methodological aspects that should be investigated in order to choose a suitable S-curve to model the performance growth of a specific technological innovation.

The findings reported in the following were first discussed in a large, unpublished research report (Erto and Vanacore 2005). These findings greatly contributed to the research activity that was begun in our department due to the pioneer work of Erto and Lanzotti (1995), and they gave rise to many research papers (e.g., Erto 1997; Vanacore 2005, 2007; D'Avino and Erto 2006, 2007; Erto and Vanacore 2008) and some theses (Di Lorenzo 1996; Medagli 2004; D'Avino 2008).

14.2 A Comparison Among Different S Curve Models

Several mathematical functions have been proposed as S-curve models in the literature. Among the most popular ones, these include the logistic, the Gompertz, the log–logistic, the Richards and the Weibull-type functions.

The main mathematical features of the models under review will be briefly illustrated here with the aid of the formulae reported in Table 14.1. Note that t represents the effort devoted to developing the technology, or an appropriate function of the time (under the assumption that the effort strictly depends on the time); $I(t)$ represents technological performance as a function of t; I_0 is a known value representing the performance level at the beginning of the innovation process; I_{lim} is the asymptotic value of the performance level that can be reached using the adopted technol-

Table 14.1 Main mathematical features of the models under review

Model	Formula	Technological performance at $t = 0$	Coordinates of the inflection point $t^*, I(t^*)$
Logistic (Verhulst 1838)	$I(t) = \dfrac{I_{\lim}}{1 + e^{\alpha - kt}}$	$\dfrac{I_{\lim}}{1 + e^{\alpha}}$	$t^* = \dfrac{\alpha}{k}$; $\quad I(t^*) = \dfrac{I_{\lim}}{2}$
Gompertz (Gompertz 1825)	$I(t) = I_{\lim} e^{-e^{\alpha - kt}}$	$\dfrac{I_{\lim}}{e^{e^{\alpha}}}$	$t^* = \dfrac{\alpha}{k}$; $\quad I(t^*) = \dfrac{I_{\lim}}{e}$
Log–logistic (Tanner 1978)	$I(t) = \dfrac{I_{\lim}}{1 + e^{\alpha - k \ln(t)}}$	0	$t^* = \left(\dfrac{e^{\alpha}(k-1)}{k+1} \right)^{1/k}$; $\quad I(t^*) = I_{\lim} \left(1 + \dfrac{k+1}{k-1} \right)^{-1}$
Richards (Richards 1959)	$I(t) = \dfrac{I_{\lim}}{\left(1 + e^{\alpha - kt}\right)^{1/s}}$	$\dfrac{I_{\lim}}{\left(1 + e^{\alpha}\right)^{1/s}}$	$t^* = \dfrac{\alpha}{k} - \dfrac{\ln s}{k}$; $\quad I(t^*) = \dfrac{I_{\lim}}{(1+s)^{1/s}}$
Generalized Weibull (Yang et al. 1978; Erto et al. 1995)	$I(t) = I_0 + \left[1 - e^{-kt^s} \right]$ $\times (I_{\lim} - I_0)$	I_0	$t^* = \left(\dfrac{s-1}{ks} \right)^{1/s}$; $\quad I(t^*) = I_0 + \left(1 - e^{(1-s)/s} \right)$ $\times (I_{\lim} - I_0)$

ogy (sometimes this is known, and in other cases it has to be estimated); α and k are scale parameters, whereas s is a shape parameter.

The coordinates of the inflection point in Table 14.1 suggest a preliminary classification of the reviewed model functions into two main categories: inflexible S-curve models (logistic and Gompertz) and flexible ones, which are classified according to whether the ordinate of the inflection point depends on the model's parameters or not.

It is obvious that using an inflexible S curve to model the performance growth of a technological innovation is risky. Indeed, the effects of the factors that influence the performance level of a technology are highly variable, and so it is unreasonable to assume that all technologies reach their maximum growth rate at a fixed distance from $I_{\lim}$.

The logistic and the Gompertz models are far and away the most commonly used S-curve models. They were originally applied to population growth processes. The logistic model is based on the hypothesis that, at the beginning, the growth rate increases exponentially, and then, as growth reaches its saturation limit, the growth rate decreases towards zero. Unlike the logistic curve, the Gompertz model does not show initial exponential growth, and it is not symmetric with respect to its inflection point, which is reached when 37% of the growth saturation limit has been attained. Both the log–logistic model and the Richards one originate from attempts to obtain a flexible S-curve model. The log–logistic model is obtained directly from the logistic one by substituting the variable t with the function $\ln t$.

The flexibility of the Richards curve is obtained at the cost of greater computational complexity. In fact, the function associated with this model is obtained by adding a third parameter, s, to the original formula for the logistic curve. The ordinate of the inflection point depends on the value of this parameter. For $s = 1$, the inflection point of the Richards model coincides with the inflection point of the logistic one, $I(t) = I_{\lim}/2$; for $s > 1$ (for $s < 1$), the inflection point of the Richards curve is obtained at a value $I(t) > I_{\lim}/2$ (at a value $I(t) < I_{\lim}/2$); finally, as s tends towards zero, the Richards growth curve approaches the Gompertz one.

Unlike the above growth models, the Weibull-type S curve derives from a morphological analogy rather than theoretical remarks. In fact, the idea of applying the Weibull probability function to growth analysis was suggested by its flexibility and the analogy between growth rate curves and probability density curves.

14.3 Criteria Used to Compare the Models

The comparison among the S curve models in Table 14.1 was carried out by considering the following six criteria:

1. Flexibility
2. Consistency with the dynamics of technological performance growth
3. Sustainability of the assumptions of linearity
4. Stability of parameter estimates versus different assumptions about the error term
5. Goodness of fit
6. Computational simplicity

We have already discussed flexibility. In order to ascertain the other comparison criteria listed above, we analyze model/data set combinations, since a specific set of observed data in conjunction with a specified model determines its behavior (Ratkowsky 1990). In order to make the results of the analysis as general as possible, three widely representative data sets have been chosen, relating to the technologies of piston aero-engines, jet aero-engines and digital signal processors (DSP).

The first two data sets are reported in Lee et al. (1988); the third data set (available from www.ti.com) refers specifically to the class of 32-bit floating DSPs introduced by Texas Instruments from 1988 to 2005.

The technological performance indicator for both the piston aero-engines and the jet aero-engines is the engine power at take-off. For the DSP data set, the technological performance is measured by its processing efficiency (proportional to the inverse of the cycle time), as suggested in Nieto et al. (1998). For all three data sets, $I_{\lim}$ is estimated to be a little bit greater than the last observed level of technological performance, since all three technologies are mature.

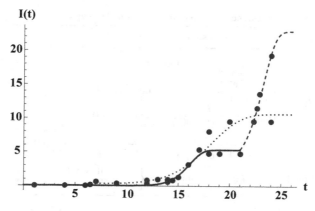

Fig. 14.1 Evolution of technological performance for DSP technologies

In Fig. 14.1, the performance growth of Texas Instruments' DSP technology is shown graphically using three S-curves. The plain curve represents the growth in the processing efficiency of the first generation of 16-bit fixed DSPs (introduced from 1982 to 1999); the dashed curve represents the growth in the processing efficiency of the second generation of 16-bit fixed DSPs (introduced from 2001 to 2005); finally, the dotted curve represents the growth in the processing efficiency of the 32-bit floating DSPs (introduced from 1988 to 2005).

14.3.1 Consistency of the Properties of the Model with the Dynamics of Technological Performance Growth

This criteria was assessed by means of a growth index called "force of obsolescence," and was defined as follows (Erto and Lanzotti 1995):

$$r(t) = \frac{dI(t)/dt}{I_{\lim} - I(t)} = \frac{i(t)}{I_{\lim} - I(t)} \tag{14.1}$$

where $i(t)$ gives, for each t, the growth rate of the obsolescence; that is, the "speed" at which the technological performance level approaches its asymptotic value $I_{\lim}$.

Unlike other growth indices proposed in the literature (Fitzhugh and Taylor 1971), $r(t)$ measures the growth rate taking the difference $I_{\lim} - I(t)$ (i.e., the residual improvement until obsolescence) as unity instead of $I_{\lim}$.

According to the common dynamics of technological performance growth, the probability that a technology will become obsolete—and thus stop being used, because it is replaced by a new and more competitive technology—increases with time. This means that the force of obsolescence in (14.1) associated with a consistent model of technological performance growth must tend to infinity as t tends to

Table 14.2 Force of obsolescence of each of the models under review

Model	Force of obsolescence
Logistic	$r(t) = \dfrac{k}{1 + e^{\alpha - kt}}; \quad \lim\limits_{t \to +\infty} r(t) = k$
Gompertz	$r(t) = \dfrac{k e^{-(e^{\alpha - kt} - \alpha + kt)}}{1 - e^{-e^{\alpha - kt}}}; \quad \lim\limits_{t \to +\infty} r(t) = k$
Log–logistic	$r(t) = \dfrac{k}{t\left(1 + e^{\alpha - k\ln(t)}\right)}; \quad \lim\limits_{t \to +\infty} r(t) = 0$
Richards	$r(t) = \dfrac{k e^{\alpha - kt}}{s\left(1 + e^{\alpha - kt}\right)\left[\left(1 + e^{\alpha - kt}\right)^{1/s} - 1\right]}; \quad \lim\limits_{t \to +\infty} r(t) = k$
Generalized Weibull	$r(t) = k s t^{s-1}; \quad \lim\limits_{t \to +\infty} r(t) = \begin{cases} +\infty & \text{if } s > 1 \\ k & \text{if } s = 1 \end{cases}$

infinity. The force of obsolescence of each model under review was formulated and is reported in Table 14.2.

Only the Generalized Weibull model shows a force of obsolescence that tends to infinity as t tends to infinity, on the condition that the parameter s is greater than one. The force of obsolescence of the log–logistic curve tends to zero as t tends to infinity; this implies that we are assuming (inconsistently) that the growth in technological obsolescence reaches a stopping point. The remaining models show a force of obsolescence that tends to a constant k as t tends to infinity, and thus they (inconsistently) imply that the growth towards obsolescence settles at a certain value. In other words, the ageing of the technology becomes independent of its current degree of obsolescence. Obviously, this is not the common case for technologies.

14.3.2 The Stability of Least Squares Estimates Versus Different Assumptions About the Error Term

In a situation where it is difficult or impossible to guarantee the assumed behavior (additive or multiplicative) of the stochastic term, it is clearly desirable for the parameter estimates of a model to remain stable under both assumptions. The stability of the parameter estimates has been evaluated by comparing the parameter estimates obtained under the assumption of additive error with the ones calculated under the assumption of multiplicative error, as suggested in Ratkowsky (1990). The least square (LS) estimates of the parameters and the residual variance, σ^2, for each model/data set combination are reported in Table 14.3, and the plots of the S-curve models estimated under the assumption of additive error are shown in Fig. 14.2. The results in Table 14.3 highlight that only the generalized Weibull model shows stable LS estimates when combined with the three considered data sets. All of the other models reviewed here, when combined with the DSP data set, exhibit unstable LS estimates. The logistic model and the Gompertz one, when used in combination with the two aero-engine data sets, present relatively stable LS estimates. The log–

logistic model and the Richards one show the worst results. Moreover, the LS parameter estimates obtained for the combination Richards model/DSP data set under the assumption of multiplicative error are not reported (n.r.), since they are affected by significant numerical errors.

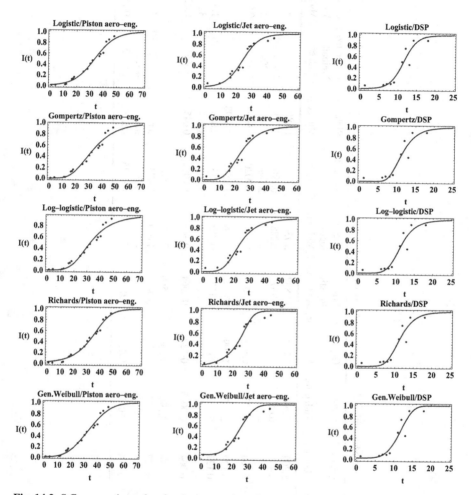

Fig. 14.2 *S*-Curves estimated under the assumption of additive error terms

Table 14.3 LS estimates for S-curve models with additive and multiplicative errors

		Logistic Add. error	Logistic Mult. error	Gompertz Add. error	Gompertz Mult. error	Log–logistic Add. error	Log–logistic Mult. error	Richards Add. error	Richards Mult. error	Generalized Weibull Add. error	Generalized Weibull Mult. error
Piston aero-engines	α	4.341	5.209	2.298	1.728	13.832	5.730	7.766	1.586	–	–
	k	0.129	0.495	0.082	0.058	3.967	2.228	0.190	0.271	1.91×10^{-5}	2.38×10^{-5}
	s	–	–	–	–	–	–	2.149	0.310	2.989	2.937
	σ^2	0.0021	0.068	0.0034	0.133	0.0037	0.170	0.0020	0.043	0.0021	0.071
Jet aero-engines	α	4.366	3.179	2.462	1.199	13.132	3.186	11.541	27.198	–	–
	k	0.184	0.127	0.122	0.059	4.177	0.919	0.383	0.848	4.62×10^{-6}	1.32×10^{-5}
	s	–	–	–	–	–	–	3.402	8.904	3.725	3.390
	σ^2	0.0051	0.064	0.0065	0.120	0.0064	0.308	0.0050	0.045	0.0051	0.029
32-Bit floating DSP	α	6.288	3.778	3.796	1.336	14.560	3.436	2.692	n.r.	–	–
	k	0.550	0.277	0.370	0.107	6.007	1.090	0.404	n.r.	6.362×10^{-6}	4.185×10^{-6}
	s	–	–	–	–	–	–	0.207	n.r.	4.728	4.837
	σ^2	0.0158	0.314	0.0153	0.466	0.0151	0.732	0.0175	n.r.	0.0163	0.157

14.3.3 Sustainability of the Assumption of Linearity

For nonlinear models, such as the S-curve, the LS estimators have unknown properties for samples of finite size. A nonlinear model whose properties are close to being what one expects from a linear one is preferable to a model whose behavior is far from linear. In fact, in this case: (a) it is easier to obtain the LS estimates; (b) interpretation is relatively straightforward; (c) the predicted values of the response variable, $I(t)$, are almost correct; (d) the shape of the joint confidence region for the parameters is nearly ellipsoidal (Ratkowsky 1990).

The approach used in order to assess the nonlinear behavior of the S-curves reported in Table 14.1 is the one proposed in Bates and Watts (1998), which quantifies, using two indices, the curvature of the parameter lines on the solution locus (*intrinsic nonlinearity*, IN) and their lack of parallelism and unequal spacing (*parameter-effects nonlinearity*, PE). Unlike the intrinsic nonlinearity, the parameter-effects nonlinearity can often be reduced by a suitable reparameterization. The curvature measures of Bates and Watts have been calculated for each reviewed model/data set combination. The statistical significance of each of the IN and PE indices is assessed by comparing its value with $1/(2\sqrt{F})$, where F indicates the $1 - \gamma$ percentile of the F distribution with p and $n - p$ degrees of freedom (n and p are the number of observations and parameters, respectively). The results are reported in Table 14.4, with the critical values (for $\gamma = 0.05$) given in parentheses. The values in bold indicate the model/data set combinations for which the specific linearity assumption is rejected.

The results show that:

- All of the reviewed models, when combined with the DSP data set, are characterized by significant nonlinearity.
- The Richards model is characterized by significant nonlinearity for all of the data sets considered here.

Table 14.4 Curvature measures of Bates and Watts and critical values ($\gamma = 0.05$) for each model/data set combination

	Curvature measures	Logistic	Gompertz	Log–logistic	Richards	Generalized Weibull
Piston aero-engines	IN	0.0902 (0.2606)	0.1680 (0.2606)	0.1582 (0.2606)	**0.3391** **(0.2734)**	0.0929 (0.2606)
	PE	0.1199 (0.2606)	0.1785 (0.2606)	0.1666 (0.2606)	**2.817** **(0.2734)**	**19.464** **(0.2606)**
Jet aero-engines	IN	0.1417 (0.2537)	0.1883 (0.2537)	0.1313 (0.2537)	**1.095** **(0.2640)**	0.2217 (0.2537)
	PE	0.1641 (0.2537)	0.1636 (0.2537)	0.1844 (0.2537)	**6.0034** **(0.2640)**	**43.73** **(0.2537)**
32-Bit floating DSP	IN	**0.2975** **(0.2368)**	**0.4669** **(0.2368)**	**0.3321** **(0.2368)**	**0.9434** **(0.2398)**	**0.4223** **(0.2368)**
	PE	**0.4895** **(0.2368)**	**0.5042** **(0.2368)**	**0.5207** **(0.2368)**	**393.57** **(0.2398)**	**85.21** **(0.2368)**

Table 14.5 Box's percentage bias index for each model/data set combination

		Logistic	Gompertz	Log–logistic	Richards	Generalized Weibull
Piston	α	0.3681	0.7343	0.7544	6.550	–
aero-engines	k	0.3590	0.6363	0.7432	6.190	27.27
	s	–	–	–	8.131	0.352
Jet	α	0.9443	1.105	1.185	53.84	–
aero-engines	k	0.9390	1.023	1.766	51.49	116.93
	s	–	–	–	60.65	1.158
32-Bit	α	5.425	5.607	5.811	–724.333	–
floating DSP	k	5.321	5.144	5.731	36.023	488.137
	s	–	–	–	187.9	6.015

- The logistic, the Gompertz, and the log–logistic models show close-to-linear behavior when combined with the two aero-engine data sets
- The generalized Weibull model, when combined with the two aero-engine data sets, is characterized by a significant PE index.

A significant PE index means that at least one parameter in the model exhibits behavior that is far from being linear. The percentage bias (i.e., the bias expressed as a percentage of the LS estimate) is a useful quantity for identifying the parameter(s) responsible for high value(s) of the PE index. A good rule of thumb is that an absolute percentage bias that is higher than 1% indicates nonlinear behavior. Box's percentage bias index (Ratkowsky 1990) has been calculated for each parameter of each model/data set combination, and these values are reported in Table 14.5.

The results in Table 14.5 confirm those shown in Table 14.4. Moreover, for the generalized Weibull model, the Box percentage bias index values suggest that the high values of PE index are mainly due to the parameter k, and so most of the nonlinearity can be removed by seeking a good reparameterization for k.

14.3.4 Seeking a Good Reparameterization for the Generalized Weibull Model

In order to obtain suggestions for a good reparameterization of k for the generalized Weibull model, a simulation study was carried out (Erto and Vanacore 2005). For each of the first two applicative examples, 1000 pseudo-random data sets were considered (generating the stochastic terms from the normal model). The histograms of the LS estimates for $\hat{k}$ obtained in relation to the two aero-engine data sets are reported in Fig. 14.3.

The shapes of the two histograms suggest that the nonlinear behavior of the generalized Weibull model can be reduced by using an exponential reparameterization. In particular, since the LS estimates $\hat{k}$ for each data set are lower than one, the chosen reparameterization is a negative exponential, $\exp(-k')$, which provides positive val-

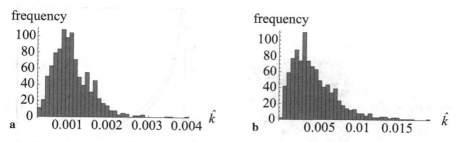

Fig. 14.3a,b Histograms of $\hat{k}$ for: **a** piston aero-engines data set; **b** jet aero-engines data set

ues of $\hat{k}'$ where $\hat{k}' = \ln(1/\hat{k})$. Thus, the reparameterized generalized Weibull model could be:

$$I(t) = I_0 + (1 - e^{-e^{-k't^s}})(I_{\text{lim}} - I_0) . \tag{14.2}$$

The percentage bias and the PE index for all combinations of the new model function (14.2) with the three data sets have been calculated and compared with the ones relating to the original model function (reported in Table 14.1). The results are reported in Table 14.6.

As a result of reparameterization, the values of the PE index as well as the percentage biases are substantially decreased and the new model function in (14.2) is much closer to linearity than the original one.

To confirm the reduction in nonlinearity resulting from the reparameterization, the error sum of squares (SSE) function and the approximate (0.90, 0.95, 0.99)% confidence contours for the parameters of both model functions after the functions are applied to the piston aero-engines data set are shown in Fig. 14.4 (note that the plots obtained for the the jet aero-engines data set are absolutely analogous to these plots).

The closer the model is to being linear, the closer the contours are to being elliptical. Thus, the change from the asymmetric "banana-shaped" contours of the original generalized Weibull model (Fig. 14.4b) to the elliptical contours of the reparameterized generalized Weibull model (Fig. 14.4d) demonstrates the reduction in nonlinear behavior (Draper and Smith 1981).

Table 14.6 Percentage biases and PE indices for the parameter k in the generalized Weibull model, and for the parameter k' in the reparameterized generalized Weibull model (critical values for the PE index are shown in parentheses)

		Generalized Weibull k	Reparameterized generalized Weibull k'
Piston aero-engines	Percentage bias	**27.27**	0.3505
	PE	**19.464 (0.2606)**	0.1157 (0.2606)
Jet aero-engines	Percentage bias	**116.93**	1.147
	PE	**43.73 (0.2537)**	0.2047 (0.2537)

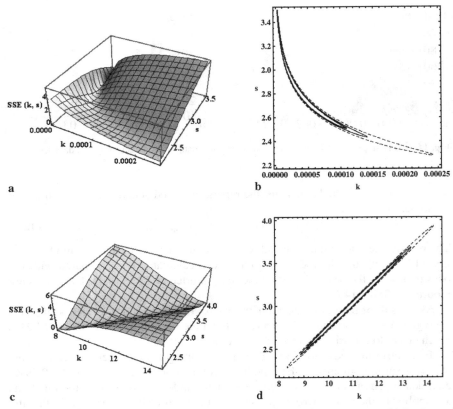

Fig. 14.4 Error sum of squares function (**a**) and approximate confidence contours (**b**) for the parameters of the generalized Weibull model to the piston aero-engines data set; error sum of squares function (**c**) and approximate confidence contours (**d**) of the reparameterized generalized Weibull model after the model has been applied to the piston aero-engines data set

14.3.5 The Goodness of Fit of Each Model

The goodness of fit of each of the models compared here was evaluated via the mean absolute percentage error (MAPE) (Carrillo and González 2002). This index is defined as follows:

$$\text{MAPE} = \frac{1}{n} \sum_{t=1}^{n} \left| \frac{I(t)^s - I(t)^o}{I(t)^o} \right| \times 100 \qquad (14.3)$$

where $I(t)^s$ is the technological performance estimated with the model function, and $I(t)^o$ is its observed value. The calculated MAPEs for each model/data set combination are reported in Table 14.7.

Table 14.7 MAPE for each model/data set combination

	Logistic	Gompertz	Log–logistic	Richards	Reparameterized generalized Weibull
Piston aero-engines	18.88	25.87	25.81	16.00	12.73
Jet aero-engines	33.23	25.34	29.30	60.26	18.59
32-Bit floating DSP	37.08	43.07	40.80	41.69	31.58

Obviously, it is not possible to state which model provides the best fit to any such data, but it is worth underlining that the reparameterized generalized Weibull gives the best fit for all three data sets examined here.

14.4 Conclusions

This chapter has provided a critical review of some of the most popular S-curve models. Six criteria were used to compare the models under review, carefully considering the dynamics of the technological innovation process as well as management needs.

An overview of the results of the comparisons is given in Table 14.8, where the crosses indicate properties held by a specific model.

Clearly, we cannot generalize these results for the three examples used here to any data set. However, it is worth noting that the reparameterized generalized Weibull is the only S-curve model reviewed in this chapter that has all of the properties that are desirable when modeling the performance growth of technological innovation.

Table 14.8 Properties held by the models under comparison

	Logistic	Gompertz	Log–logistic	Richards	Reparameterized generalized Weibull
Consistency with technological growth dynamics					X
Computational simplicity	X	X	X		X
Flexibility			X	X	X
Stability	X	X			X
Sustainability of the assumption of linearity	X	X	X		X
Goodness of fit					X

References

Asthana, P.: Jumping the technology S-curve. IEEE Spectrum. June, 49–55 (1995)

Bates, D.M., Watts, G.W.: Nonlinear Regression Analysis and its Applications. Wiley, New York (1998)

Carrillo, M., González, J.M.: A new approach to modelling sigmoidal curves. Technol. Forecast. Soc. **69**, 233–241 (2002)

D'Avino, D.: Stochastic performance modelling and management for technological systems. Thesis, Department of Aerospace Engineering, University of Naples Federico II, Naples (2008)

D'Avino, D., Erto, P.: Measuring ISO 9000 effects on a company quality performance. In: Proc. 6th Annual ENBIS Conf., Wroclaw, Poland, 18–20 Sept. (2006)

D'Avino, D., Erto, P.: A piecewise regression model for identifying the strategic inflection point for organizational change. In: Proc. 10th QMOD Conf., Helsingborg, Sweden, 18–20 June (2007)

Di Lorenzo, D.: Misure di non linearità e test di ipotesi per modelli statistici di innovazione tecnologica. Thesis, Department of Aerospace Engineering, University of Naples Federico II, Naples (1996)

Draper, N.R., Smith, H.: Applied Regression Analysis, 2nd edn. Wiley, New York (1981)

Erto, P.: Modelling of the technological innovation process. In: Conf. ISO Series 9000 Standards, Statistical Methods, Problems of Certification, Nizhny Novgorod, Russia (1997)

Erto, P., Lanzotti A.: Some tools to control the technological innovation process. Proceedings 1st World Conference for Total Quality Management, Sheffield, pp. 412–415, 10–12 April (1995)

Erto, P., Vanacore A.: A critical review of a flexible S curve model for the technological innovation process (unpublished research report). Department of Aerospace Engineering, University of Naples Federico II, Naples (2005)

Erto, P., Vanacore, A.: An adaptive approach to growth modelling (unpublished research paper) (2008)

Fitzhugh, H.A., Taylor, St.C.S.: Genetic analysis of degree of maturity. J. Anim. Sci. **33**, 717–725 (1971)

Gompertz, B.: On the nature of the law of human mortality, and on a new mode of determining the value of life contingencies. Phil. Trans. Roy. Soc. **115**, 513–585 (1825)

Lee, T.H., Nakicenovic, N.: Technology life cycle and business decision. Int. J. Manag. **3**, 411–426 (1988)

Medagli, F.A.: Modello dei fenomeni indotti da un'innovazione di prodotto. Thesis, Department of Aerospace Engineering, University of Naples Federico II, Naples (2004)

Nieto, M., Lopéz, F., Cruz, F.: Performance analysis of technology using the S curve model: the case of digital signal processing (DSP) technologies. Technovation **18**, 439–457 (1998)

Ratkowsky, D.A.: Handbook of Nonlinear Regression Models. Marcel Dekker, Inc., New York (1990)

Richards, F.J.: A flexible growth function for empirical use. J. Exp. Bot. **10**, 290–300 (1959)

Tanner, J.C.: Long term forecasting of vehicle ownership and road traffic. J. Roy. Stat. Soc. A **141**, 14–63 (1978)

Vanacore, A.: Managing the technological innovation process via a flexible S curve model. In: Proc. 8th QMOD Conf., Int. Conf. on Quality Management for Organisational and Regional Development, Palermo, Italy, pp. 415–422, 29 June–1 July (2005)

Vanacore, A.: QIC analysis: a practical tool to manage continuous quality improvement. In: Proc. 10th QMOD Conf., Int. Conf. on Quality Management for Organisational and Regional Development, Helsingborg, Sweden, 18–20 June (2007)

Verhulst, P.F.: Notice sur la loi que la population suit dans son accroissement. Corr. Math. Phys. **X**, 113–121 (1838)

Yang, R.C., Kozak, A., Smith, J.H.G.: The potential of Weibull-type functions as flexible growth curves. Can. J. For. Res. **8**, 424–431 (1978)

Index